T0310455

Digital Holography for MEMS and Microsystem Metrology

Microsystem and Nanotechnology Series

Series Editors – Ron Pethig and Horacio Dante Espinosa

Digital Holography for MEMS and Microsystem Metrology	Asundi	July 2011
Multiscale Analysis of Deformation and Failure of Materials	Fan	December 2010
Microfluidic Technology and Applications	Koch et al.	November 2000
AC Electrokinetics: Colloids and Nanoparticles	Morgan and Green	January 2003
Fluid Properties at Nano/Meso Scale	Dyson et al.	September 2008
Introduction to Microsystem Technology	Gerlach	March 2008

Digital Holography for MEMS and Microsystem Metrology

Edited by

Anand Asundi

School of Mechanical and Aerospace Engineering
Nanyang Technological University, Singapore

A John Wiley & Sons, Ltd., Publication

This edition first published 2011

Registered office
John Wiley & Sons Ltd, The Atrium, Southern Gate, Chichester, West Sussex, PO19 8SQ, United Kingdom

For details of our global editorial offices, for customer services and for information about how to apply for permission to reuse the copyright material in this book please see our website at www.wiley.com.

Library of Congress Cataloguing-in-Publication Data

Digital holography for MEMS and microsystem metrology / edited by Anand Asundi.
p. cm.
Includes bibliographical references and index.
ISBN 978-0-470-97869-6 (cloth)
1. Microelectromechanical systems–Measurement. 2. Microelectronics–Measurement. 3. Holographic testing. 4. Image processing–Digital techniques. I. Asundi, Anand.
TK7875.D54 2011
621.381–dc22

2011009635

A catalogue record for this book is available from the British Library.

Print ISBN: 9780470978696
ePDF ISBN: 9781119997306
oBook ISBN: 9781119997290
ePub ISBN: 9781119972785
Mobi ISBN: 9781119972792

Set in 10/12pt Palatino-Roman by Thomson Digital, Noida, India

To my wife Radha (Champa) Asundi for all her patience and perseverance

Contents

About the Editor

Anand Asundi (安顺泰) graduated from the Indian Institute of Technology, Bombay, with a BTech (Civil Eng.) and an MTech (Aeronautical Eng.). Subsequently he received his PhD from the State University of New York at Stony Brook. Following a brief tenure at the Virginia Polytechnic Institute and State University, he was with the University of Hong Kong from 1983 to 1996 as Professor in the Department of Mechanical Engineering. He is currently Professor in the School of Mechanical and Aerospace Engineering at the Nanyang Technological University in Singapore. His teaching area is in Solid Mechanics and his research interests are in optical methods in mechanics, including micro, nano and bio mechanics, on-line structural health monitoring and fiber optic bio-chemical sensors. He has published extensively and presented invited seminars/talks at various institutions and at international conferences. He is the editor of *Optics and Lasers in Engineering* and a Fellow of the Society of Photo-Optical Instrumental Engineers (SPIE), the International Society of Optical Engineers and Institution of Engineers, Singapore, and a member of the Optical Society of America. He is the founding chair of the Optics and Photonics Society of Singapore, the Asian Committee on Experimental Mechanics and the Asia Pacific Committee on Smart Materials and Nanotechnology. He has organized numerous conferences and served on the Membership, Scholarship/Awards and Presidential Asian Advisory Committees of the SPIE and on the Board of Directors of the SPIE.

Contributors

Vijay Raj Singh received his Master of Technology (MTech) degree in Applied Optics from the Indian Institute of Technology (IIT) Delhi, India, in 2003, and his PhD in Optical Science and Engineering from Nanyang Technological University (NTU), Singapore, in 2008. His research interests include digital holography, image processing, optical metrology, microscopy, and 2D and 3D imaging. From 2007–2010, he worked at AEM Singapore Pte Ltd and Nanyang Technological University, Singapore, and his research work focused on the development of digital holographic microscopes as tabletop and handheld systems for MEMS characterization and 3D imaging of live bio-cells. Currently he is working for the Singapore–MIT Alliance for Research and Technology (SMART) Center, Singapore, and working on image processing methods for biomedical applications. He has one US patent pending, and has published 12 peer-reviewed research articles in international journals and 20 papers in international conference proceedings. He is a member of the SPIE (the Society of Photo-Optical Instrumental Engineers) and the OSA (the Optical Society of America). He was one of the founders of the Singapore student chapter of the SPIE and worked as the President of the chapter from 2005–2007. He also served as a committee member of the Optical and Photonics Society of Singapore (formerly the SPIE Singapore chapter) from 2007–2010.

Qu Weijuan received her MS in Optics from Northwestern Polytechnical University in 2004 and PhD in Optics Engineering from the Shanghai Institute of Optics and Fine Mechanics, the Chinese Academy of Sciences, in 2007. She then spent two years as a research staff at Nanyang Technological University. Presently, she is an optics engineer in the Center of Innovation, Ngee Ann Polytechnic. Dr. Qu has been working in digital holography for about 10 years. Her research interests include theoretical and experimental technique development and the application of digital holographic microscopy, live cell imaging, and micro-optics characterization.

Taslima Khanam obtained her BSc in Chemical Engineering from Bangladesh University of Engineering and Technology (BUET) in 2006. She submitted her doctoral thesis to the division of Chemical and Biomolecular Engineering, Nanyang Technological University (NTU) in Jan. 2011 (at the time of contributing to this book). During her PhD studies she won the best student paper award at the 9th International Symposium on Laser Metrology, Singapore, in 2008. Her work also received the best paper award in the 2009 AIChE (American Institute of Chemical Engineers) annual meeting (Presenter: Asst. Prof. Vinay Kariwala, NTU). Her research focus is the development of optics-based tools for on-line measurements of particle size and shape for the application of particulate processes.

Caojin Yuan received her BSc in Optoelectronics and MSc in Physical Electronics in 2002 and 2005, respectively, from Kunming University of Science and Technology, China and a PhD in Optical Engineering from Nankai University, China in 2008. She is currently a research fellow at Stuttgart University, Germany and is supported by the Alexander von Humboldt Foundation. Her research interests include digital holography, microscopic imaging and optical information processing.

Hongchen Zhai received his PhD at the Universität Münster, Germany, in 1990, and then spent three years in the Laboratoire Charles Fabry, Institute of Theoretical and Applied Optics, National Scientific Research Center of France. He is now a professor and the Academic Committee Member at Nankai University, China, and the Deputy Director of the Committee of Holography and Optical Information Processing, of the China Optical Society. His research interests include pulsed digital holography and optical information processing.

Yu Yingjie obtained her bachelor's degree with a major in Precision Instruments from Harbin University of Science and Technology, China in 1991, and a master's degree and doctoral degree in 1996 and 1998, respectively, with a major in Precision Instrument and Mechanics from Harbin Institute of Technology, China. From 1991 to 1998, she also worked in the precision instruments laboratory of the Harbin Institute of Technology, as a teacher of experimental courses and did some testing work. From 1999 to the present, she has been working in the Department of Precision Mechanical Engineering at Shanghai University, China. In 2005, she gained her professorship. Her research area is applied optics and metrology and her research interest focuses on digital interferometry, digital holography and electronic speckle interferometry. Her main works include the system and software designing of phase-shifting interferometer by PZT and via wavelength tuning, sub-aperture stitching interferometry, digital micro-holography and its application in biology, digital holographic tomography, computer-generated holograms and three-dimensional holographic displays, electronic speckle interferometry and three-dimensional deformation measurement, designing and testing optical ultra-small probes for biomedical imaging. In recent years, she has been responsible for more than 10 research projects, has published more than 60 papers and has been granted rights to five patents.

Jianlin Zhao received his BS degree in Applied Physics and his MS degree in Solid Mechanics from Northwestern Polytechnical University (NPU), China, in 1981 and 1987, respectively, and his PhD in Optics from Xi'an Institute of Optics and Fine Mechanics, the Chinese Academy of Science, China in 1998. Currently, he is a Physical Professor in the Department of Applied Physics, School of Science, NPU. He is also one of the council members of the Chinese Optical Society (COS), and one of the vice-chairmen of three specialty committees (Holography and Optical Information Processing, Optical Education, and High Speed Photography and Photonics), COS. He has published two optical textbooks (*Optics*, and *Advanced Optics*, both in Chinese) and over 270 research articles in journals and papers in international conference proceedings. He is also one of the authors of the *Handbook of Optics* (Chapter 15: Information Optics, edited by Prof. Jingzhen Li, 2010, in Chinese). His specialization is optical engineering and his current research interests are optical information technologies (micro-nano photonics, digital optical information processing, digital holography, optical fiber sensors and applications).

Series Preface

The original concept and theory of electron holography was developed in 1947 by Dennis Gabor as a way to improve the resolution of electron microscopy, but its practical realisation in the optical form we know today had to await the development of coherent light sources (lasers) in the 1960's. Countless numbers of laboratories and photographic studios now use standardised equipment, typically consisting of a continuous wave laser, lenses and beam splitters to construct holographic images. The ability to record the three-dimensional details of an object in a single hologram often makes this the technique of choice for imaging and measurement.

Although the technology might appear to be mature, with only minor improvements achievable, major issues requiring attention do exist. For example, the photographic plates must be isolated from mechanical vibrations during their exposure. Such mechanical stability is absolutely essential because movement as small as a quarter wave-length of light during exposures can completely ruin a hologram. In some industrial environments this problem can be overcome using pulsed lasers rather than continuous wave lasers, but this adds another layer of complexity onto what is already a complicated process. Also, the wet-chemistry required to process the photographic plates can be expensive and time-consuming. The continuing trend towards miniaturization down to the micro- and nano-scales increases the challenges facing the use of holography in imaging and metrology.

In this book Professor Anand Asundi has assembled excellent contributions from experts at the forefront of developing exciting and important applications of digital holography for micro-measurements on micro-devices and MEMS structures. We learn that the recent development of digital computers and charged coupled devices provide the means to record holograms directly in digital form at video rates. Reconstruction of the images can then be performed numerically through quantitative analyses of the amplitudes and phases of the stored interference patterns. Digital technology has thus made it possible to both record and very flexibly reconstruct holograms using computers. The potential of this is very exciting!

The style of writing is pedagogical, making this book suitable for experts in the field as well as undergraduate and postgraduate students attending courses in electronic engineering, materials science, MEMS, applied physics or computing.

Ronald Pethig
Horacio D. Espinosa

Acknowledgements

This work would not have been possible but for the hard work and dedication of my research students, primarily Xu Lei who started Digital Holography at the Nanyang Technological University (NTU), Singapore. Vijay Raj Singh, Taslima Khanam, Qu Weijuan and Yan Hao have been instrumental in moving this forward. I would also like to acknowledge Sui Liansheng, Di Jianglei and Chee OiChoo who have contributed in no small way to this effort.

The work was supported by the Microfabrication Centre at NTU and research grants through the National Science Foundation and the Ministry of Education, Singapore.

Anand Asundi

Vijay Raj Singh would like to thank Prof. Anand Krishna Asundi for providing him with an opportunity to work together on digital holography for static and dynamic metrological applications for MEMS and micro-system characterization and for inviting him to write a chapter for this book. Digital holography is an exciting new method for handling of light and he believes this book will provide readers with an insight into the recent technological developments and implementation of digital holography-based techniques for MEMS and micro-systems testing. He would also like to express his gratitude to his wife for her constant encouragement.

Qu Weijuan would like to thank Prof. Anand Krishna Asundi for enabling this work to be published and all the valuable suggestion and help. She would like to thank Ms Chee Oi Choo, who provided support, read and offered comments. Above all she wants to thank her husband Zhou Jianbo and the rest of his family for their support and encouragement. Qu Weijuan also gratefully acknowledges the support of Innovation Fund grant MOE2008-IF-1-009 from the Singapore Ministry of Education and Translational Research and Development grant NRF2009NRF-TRD001-008 from the Singapore National Research Foundation.

Taslima Khanam acknowledges funding from Nanyang Technological University through AcRF Tier 1 Grant no. RG25/07. She also thanks

Dr. Arvind Rajendran, Dr. Vinay Kariwala, Dr. Emmanouil Darakis and Dr. Michel Kempkes for their valuable suggestions, support and assistance in this work.

Caojin Yuan and Hongchen Zhai gratefully acknowledge the support of the National Natural Science Foundation of China under Grant No. 60838001 and No. 60907002.

Yu Yingjie gratefully thanks Dr. Wenjing Zhou and PhD student Li Zhao of Shanghai University, China, for providing and organizing numerous useful materials. A special thank you is given to Professor Anand Asundi of Nanyang Technological University, Singapore, for his invaluable comments and advice.

Jianlin Zhao thanks the National Natural Science Foundation of China under Grant No. 60077018 and 61077008 and the Science Foundation of Aeronautics of China under Grant No. 02I53075 and 2006ZD53042 for their financial support of the research work.

Abbreviations

2D	two-dimensional
3D	three-dimensional
ADM	angular division multiplexing
ALD	axis length distribution
AlN	Aluminum nitride
ART	Algebra Reconstruction Technique
ASM	angular spectrum method
ATM	atomic force microscope
BS	beam splitter
BUET	Bangladesh University of Engineering and Technology
CCD	charge coupled device
CE	circle equivalent
CGH	computer generated holography
CLSM	confocal laser scanning microscope
CMOS	complementary metal oxide semiconductor
COS	Chinese Optical Society
CT	computer tomography
DH	digital holography
DHM	digital holographic microscopy
DHT	digital holographic tomography
DIH	digital holographic interferometry
FFT	fast Fourier transform
FT	Fresnel transform
IIT	Indian Institute of Technology
LDIHM	lens-less digital in-line holographic microscopy
LDM	long distance microscope
LDV	laser doppler vibrometer
MEMS	micro-electromechanical system
MO	microscopic objective
MOES	micro-opto-electromechanical system
MPT	microwave plasma thruster
NA	numerical aperture
NIST	National Institute of Standards and Technology
NPU	Northwestern Polytechnical University
NTU	Nanyang Technological University
OCT	optical coherence tomography
PCA	principal component analysis
PM	polarization multiplexing

PSD	particle size distribution
RDM	recording plane division multiplexing
ROC	radius of curvature
SCBS	single cube beam splitter
SEM	scanning electronic microscope
SM	spatial multiplexing
SMART	Singapore-MIT Alliance for Research and Technology
SOI	silicon on insulator
VCSEL	vertical-cavity surface-emitting laser
WLIM	white-light interference microscope
WDM	wavelength division multiplexing

1

Introduction

Anand Asundi
School of Mechanical and Aerospace Engineering, Nanyang Technological University, Singapore

An optical wave is characterized by its amplitude, frequency, phase, polarization, and direction of propagation. When a coherent optical wave is incident on any object, the reflected and/or the transmitted waves contain information about the optical and physical properties of that object. The amplitude contains information about reflectance or attenuation of the object, while the phase gives topography or thickness characteristics. Thus, both these parameters are important for the complete three-dimensional (3D) study of objects. Optical measurement techniques offer significant advantages over their counterparts for imaging and measurement applications. Remote analysis, non-contact measurement, whole field visualization, and no need for special sample preparation are the major advantages. The increasing possibilities of computer-aided data processing have led to a new revival in optical metrology. Recent technological developments and miniaturization of the test objects are creating new challenges for optical metrology, for example, to provide a convenient tool for whole field imaging and micro-systems characterization, and to provide experimental data for computer-aided engineering for fast and accurate measurements, and so on. Different optical methods are used for these measurements depending on the requirements. These methods can be divided into two broad categories, called imaging and interferometric methods, summarized in Figure 1.1.

Digital Holography for MEMS and Microsystem Metrology, First Edition. Edited by Anand Asundi.

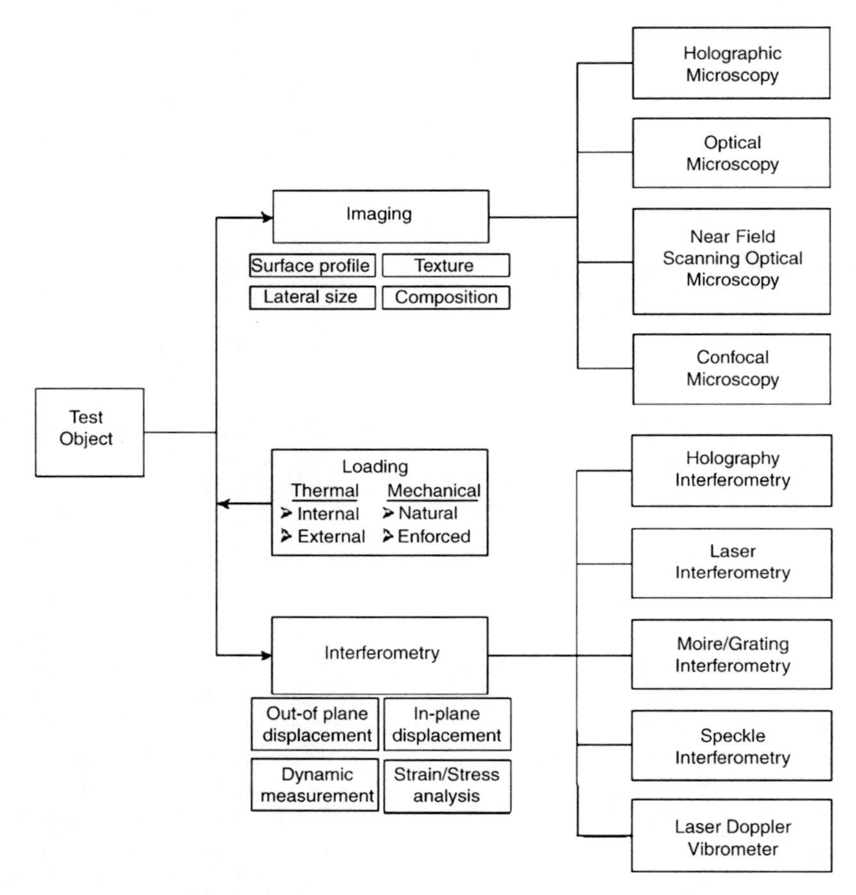

Figure 1.1 Methods of optical metrology

New challenges for the imaging and measurement processes introduced by the miniaturization of the test objects require the development of reliable advanced testing methods. Some examples are dynamic microscopic imaging (for example, micro-particles image velocimetry, micro-fluids flow analysis, and the study of biological samples), and static and dynamic measurement of micro-structures. The integration of mechanical elements, electronics, sensors and actuators on a common silicon substrate by micromachining technology constitutes a micro-electromechanical systems (MEMS). This has a wide range of applications in scientific and engineering fields. Characterization of the mechanical properties of MEMS structures at different stages of manufacturing is extremely important. The aim of this testing is to provide feedback about device behavior, system parameters, and material properties for the design and simulation processes. Also dynamic testing is needed in the final devices to test their performance and characteristics. 3D imaging and characterization of the mechanical properties of MEMS structures are a challenging task.

Various techniques have been explored to characterize MEMS devices. Thermographic techniques such infra-red radiation analysis, fluorescent micro-thermographic imaging techniques and liquid crystal methods have been used in the thermal characterization of MEMS devices. These techniques, however, have their limitations such as poor resolution, issues concerning repeatability or coating the device with different layer. A non-destructive optical technique for thermal deformation characterization has been used. Though the technique provides a spatial resolution of 0.5–1 μm, the main difficulty arises with the need to know the reflectivity coefficient of the material used. The above-mentioned techniques are useful in estimating the device temperature. To characterize the deformations in the device, different techniques have been adopted, such as a 3D surface profilometer, involving a white light interferometric scanning principle with a stroboscopic LED light source, providing a vertical displacement resolution of 3–5 nm. In-plane motion characterization of MEMS resonators could be performed using a stroboscopic scanning electron microscope imaging technique. The accuracy of the measured displacement using this technique is about 20 nm, limited mainly by the electron probe size and the digital scanning resolution. Laser doppler vibrometry is also one of the widely used MEMS characterization techniques. Frequency response of vibration amplitude of the mechanical structures, along with their vibration modes, can be obtained using a vibrometer, but it cannot provide the static deformation of the mechanical structures. Furthermore, they provide vibration information only at a single point. To analyze the vibrations of a device, the laser beam has to scan the entire structure.

Holography is an important tool for optical metrology. Dennis Gabor invented holography in 1948 as a two-step lens-less imaging process for wavefront reconstruction. The phase, amplitude, polarization, and coherence of a wave field can be stored in a hologram during recording. Since these quantities cannot be measured directly with conventional detectors, which are only sensitive to the intensity, this makes holographic technique most attractive. Holography is well established for scientific and engineering studies and has found a wide range of applications. In classical holography, photographic plates or thermoplastic films are used to record holograms in a vibration-free environment and then are optically reconstructed. The handling of these materials is time-consuming and the vibration isolation requirement makes holographic technique less popular in industrial environments. In addition, when smaller objects are studied, the complicated experimental set-up and evaluation processes during reconstruction often impose challenging problems.

With the development of digital computers and digital recording devices, that is, a charged coupled devices (CCD) and CMOS sensors, digital holography was proposed to overcome the problems of classical holography. CCD sensors provide the flexibility to record holograms directly in digital form. The reconstruction process is then performed numerically, giving a quantitative analysis of amplitude and phase of the wavefront. This offers new possibilities for a variety of applications, which in classical holography was done only qualitatively. Digital holography is an exciting new method for handling light. The capability of whole field information storage in holography and the use of computer technology for fast data processing open up a lot of possibilities to develop digital holography as a novel metrological tool. This has received increased attention in the past two decades. The main reason for this development has been the rapid improvement in storing and processing of digital information. The object wave can be numerically calculated directly from the holograms and this makes it possible to quantitatively calculate both the amplitude and phase information separately, which is not directly possible in conventional holography and other classical optical metrological methods. These features make it the perfect method for both imaging and measurement. Digital holography has the potential to address recent challenges for optical metrology brought about by miniaturization, new material and processes.

CCD sensors as holographic recording medium allow the recording of holograms at video rates and are in a format readily available for numerical reconstruction. Commercially available CCD sensors have their limitations, such as lower pixel resolution and smaller sensing area. This limited resolution restricts the angle between object and reference beams to a few degrees only. To achieve a good quality reconstruction, the digital recording of holograms has to satisfy the requirements of Nyquist sampling theorem. This implies that interference between two beams of wavelength λ intersecting at an angle θ gives fringe spacing $p = \lambda/\sin\theta$ and it must be larger than two pixels. For example, for 3° interference angle between two beams of He-Ne laser ($\lambda = 0.6328\ \mu m$), the pixel size of CCD should be $6\ \mu m$ to fulfill the sampling theorem. This corresponds to the higher end of the CCD cameras. It is interesting to note that it is the pixel size along with number of pixels of a CCD sensor that is also important. An overview of the research work focused on digital holography and its applications explored in the past decade is shown in Figure 1.2.

One of the most remarkable applications of holography is in the interferometric comparison of diffuse wavefronts. This technique uses the vast generalization of interferometry to measure displacement and surface deformation of opaque objects. It involves the double exposure method from two different states of the object. By reconstructing the hologram, the two wavefronts interfere with each other and produce a fringe pattern that describes the changes that have occurred between the two states of the object. The numerical reconstruction process in digital holography makes it possible to numerically measure the intensity and phase for both states of the object directly from the holograms. It is possible to calculate the interference phase directly

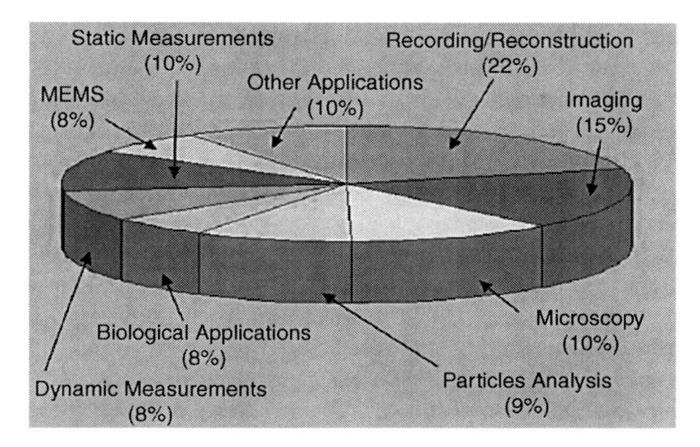

Figure 1.2 Research overview of digital holography in the past decade

from the holograms, which is the phase difference between the two object waves. Also the other methods of optical metrology, for example, shearography and speckle photography, can be numerically obtained from digitally recorded holograms. Ways of measuring micro-deformation based on the principle of digital holographic interferometry have been widely studied.

Digital holography offers new possibilities in the non-invasive measurement of MEMS devices. In digital holography, the numerical separation of amplitude and phase enables the direct determination of the modulo 2π interferometric phase without the need for any phase shifting method while at the same time the numerical amplitude reconstruction allows for lens-less imaging. Strain analysis has been described by digital holographic methods. By illuminating the object from three directions, all three surface displacement components (x, y, z) are calculated during the reconstruction process. These components can be used to find the displacement gradient or strains on the surface. Incorporation of microscopy with digital holographic interferometry improves the capability of measurements for microsystems and has been used to measure deformation with an accuracy of less than 1 nm. Use of digital holographic microscopy (DHM) in MEMS characterization has generated a lot of interest in recent years because of the accurate determination of the deformations due to the residual stress and the impact of thermal loads on deformations. DHM provides a non-contact-based and non-destructive method for static as well as dynamic characterization of MEMS devices. Dynamic digital holography has also attracted great interest in recent years. There are two approaches to this – the first is to use a pulse laser or high speed camera to record multiple frames which are then processed much like the static case. Methods such as pulse or stroboscopic digital holography require precise synchronization of the light source, specimen and the recording device. The second is to use the time average method which does not require any high speed camera or a pulsed laser. In this book, digital holographic systems that use off-axis geometry and in-line lens-less geometry are presented for micro-size object imaging and measurement (static and dynamic) applications. The in-line digital holographic system has shown better performance due to its higher space-bandwidth product which provides a larger field of view and higher imaging resolution than the off-axis set-up.

This book deals with topics related to micro-measurement with a specific focus on micro-devices and MEMS structures for which digital holography is best suited. In Chapter 2, the reflection mode of digital holography is described, including the recording using in-line, off-axis and lens-less geometries. This is followed by examples of 3D characterization of MEMS devices as well as static and dynamic testing of these components. Specific applications are highlighted in this chapter. Chapter 3 deals with transmission digital holographic configurations. Since in this case the reference and object beams need to be separately formed, some understanding of two beam interference under different conditions is reviewed to highlight the effect of phase compensation. These are then incorporated into the design of three different configurations where the phase is physically rather than numerically compensated. Applications of these are demonstrated for micro-measurements. Chapter 4 deals with the use of an in-line digital holography system for micro-particle sizing and counting both in static as well as in dynamic situations. There is interest in this in the field of crystallization studies, as particles which are both spherical as well as needle-shaped can be measured in a 3D volume. Chapter 5 contains a group of related applications starting with digital tomography (Chapter 5.1), followed by high resolution pulsed digital holography (Chapter 5.2), and finally digital holography for refractive index measurement with applications in photorefractive materials, acoustics, plasmas, turbulent flow and thermal measurements (Chapter 5.3).

2

Digital Reflection Holography and Applications

Vijay Raj Singh[1] and Anand Asundi[2]
[1]*Singapore - MIT Alliance for Research and Technology (SMART) Center, Singapore*
[2]*School of Mechanical and Aerospace Engineering, Nanyang Technological University, Singapore*

2.1 Introduction to Digital Holography and Methods

2.1.1 Holography and Digital Holography

Holography is a well-established method of optical metrology for imaging and measurement applications. It has several additional advantages over other conventional optical microscopy and interferometer methods. Amplitude and phase are the main data used to characterize any optical wave. Optical detectors such as photographic films or CCDs can record only the amplitude, and the phase information is lost in this recording process. Holography, on the other hand, records both the amplitude and phase, as proposed by Dennis Gabor [1]. When an object wave interferes with a known reference wave, the recorded interference pattern is called a hologram. The data of the object wave is encoded in the interference pattern in terms of fringe modulation, location and spacing. The object wave can be reconstructed by illuminating the

Digital Holography for MEMS and Microsystem Metrology, First Edition. Edited by Anand Asundi.

hologram with the same reference wave, as used during the recording process. The first order diffracted wave from hologram is the same as the object wave as successfully demonstrated, after the invention of lasers, by Leith and Upatniek who used an off-axis configuration to separate the object wave from the non-diffracted wave [2]. The most attractive part of holography is the recording of full 3D data about the object in a single hologram. This feature makes holography the best suited tool for imaging and measurement.

The major drawback of classical holography is the need for photographic plates, which involves a wet chemical development process that is expensive and time-consuming. Also, the vibration isolation requirement in holography is not compatible with industrial shop floor environments unless pulsed lasers are used. Moreover, the exact repositioning of the holograms for real-time measurements, the dependence of the reconstructed images and the interference fringes on the optical system, and phase-shifting techniques all make classical holography a complicated process.

Along with the recent development of digital computers and charged coupled devices (CCD), digital holography is proposed as a solution to overcome the problems of classical holography. Digital recording devices (a CCD sensor) provide flexibility to record holograms directly in digital form [3]. The reconstruction process is then performed numerically, giving quantitative analysis of the amplitude and phase of the wavefront. This offers new possibilities for a variety of applications, which in classical holography was performed only qualitatively. Digital technology has made both recording and reconstruction of holograms by computers possible. When holograms are created by numerical simulations and reconstructed optically, this is called computer-generated holography (CGH). On the other hand, when an optical hologram is recorded digitally by a CCD and is reconstructed numerically, it is called digital holography (DH). CCD sensors as the holographic recording medium allow the recording of holograms at video rates and are in a format readily available for numerical reconstruction. The computer reconstruction process is very flexible and can be used in a variety of ways to focus the image at different planes or remove the background, and so on. The process of digital holography is shown in Figure 2.1.

Digital holography is an exciting new method for handling light. The capability of whole field information storage in holography and the use of computer technology for fast data processing open up the possibility of

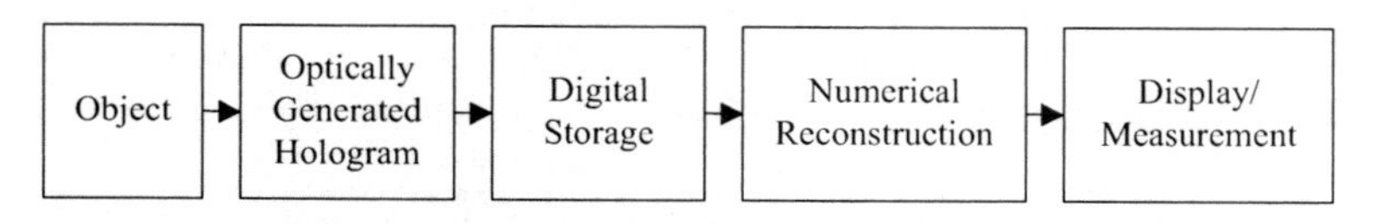

Figure 2.1 The process of digital holography

developing digital holography as a dynamic metrological tool. As an accepted method, digital holography has received increased attention and has been developed for various applications in the past few years. Digital holography has the potential to address recent challenges for optical metrology brought about by miniaturization, new materials and new processes.

Digital recording of a hologram, using CCD sensors, needs to fulfill the sampling theorem across the entire sensing area. Commercially available CCD sensors have their limitations, such as low pixel resolution and a smaller sensing area. An off-axis digital holography system utilizes only a partial area of a CCD sensor and provides a poor imaging resolution. In contrast, the in-line system helps to relax the spatial resolution requirement on CCD sensors and utilizes the full sensing area.

2.1.2 Digital Recording Mechanism

The diffraction region of the object wave, where the digital hologram is recorded, is used to classify the digital hologram. It is divided into two categories: the Fresnel hologram and the Fraunhofer hologram. The digital Fresnel holography set-up is shown in Figure 2.2. The interference of the reference beam and the object beam is recorded by a CCD sensor which is placed in the Fresnel diffraction region of the object wave. The collimated reference beam is incident normally on the CCD and θ is the offset angle of the object.

For efficient utilization of the recording sensor, it is important that the sampling theorem must be fulfilled across its entire CCD sensing area. Thus, for a certain lateral size of the object, the recording distance (between the object and the CCD sensor) should be greater than a particular minimum value. The limited resolution of CCD sensors limits the angle θ (called the interference angle) between the object and the reference beams to few degrees only. For the exact recovery of the object data during the reconstruction process, the sampling

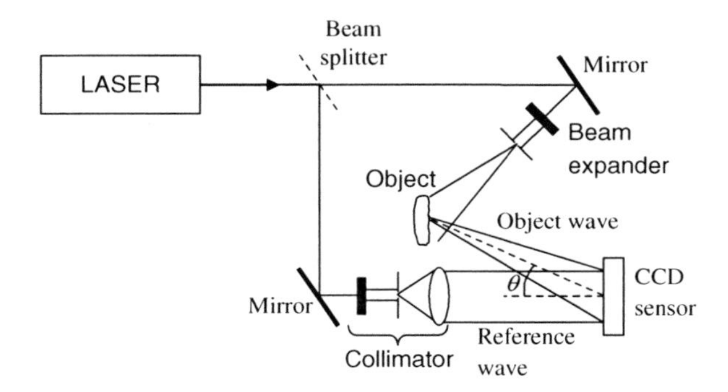

Figure 2.2 The digital Fresnel holography set-up

theorem requires that the interference fringe spacing must be larger than the size of two pixels of the CCD that is:

$$\theta < \frac{\lambda}{2\Delta_x} \tag{2.1}$$

where λ is the wavelength of light used and Δ_x is the pixel size of the CCD. The recording distance, between the object and CCD, is an important parameter to control the interference angle.

During the interference of the object and the reference, the waves propagate along the same optical axis. Consider (x, y) to be the hologram plane and the object and reference waves are denoted by $O(n_x, n_y)$ and $R(n_x, n_y)$ at the CCD plane. Here $n_x = 0, 1, \ldots N_x - 1$ and $n_y = 0, 1, \ldots N_y - 1$ are the pixel indices of the camera, and $N_x \times N_y$ is the size of the CCD sensor in pixels. The hologram is the interference of the object wave reference waves and can be written as:

$$\begin{aligned} h(n_x, n_y) &= | \ O(n_x, n_y) + R(n_x, n_y)|^2 \\ &= | \ O(n_x, n_y)|^2 + |R(n_x, n_y)|^2 + O*(n_x, n_y)R(n_x, n_y) + O(n_x, n_y)R*(n_x, n_y) \end{aligned} \tag{2.2}$$

Here $O*$ and $R*$ are the complex conjugate of O and R, respectively. The CCD, placed at the hologram plane, records the interference pattern as given in Equation 2.2. For digital recording, the sampling theorem requires that the interference fringe spacing must be larger than the size of two pixels of CCD. The recorded pattern is converted into a two-dimensional array of discrete signals by using the sampling theorem. Let the pixel size of the CCD be Δ_x and Δ_y, then the digitally sampled holograms, can be written as:

$$H(n_x, n_y) = \left[h(n_x, n_y) \otimes rect\left(\frac{x}{\alpha\Delta_x}, \frac{y}{\beta\Delta_y} \right) \right] \times rect\left(\frac{x}{N_x\Delta_x}, \frac{y}{N_y\Delta_y} \right) comb\left(\frac{x}{\Delta_x}, \frac{y}{\Delta_y} \right) \tag{2.3}$$

where $\otimes$ represents the two-dimensional convolution and $(\alpha, \beta) \in [0, 1]$ are the fill factors of the CCD pixels.

2.1.3 Numerical Reconstruction Methods

The reconstruction of a hologram is a diffraction process. The diffraction geometry is shown in Figure 2.3.

Numerical reconstruction of a hologram is performed using the Fresnel diffraction theory. The hologram $H(n_x, n_y)$, recorded at the (ξ, η) plane, is multiplied by the numerical reconstruction wave $R(n_x, n_y)$ and the numerically

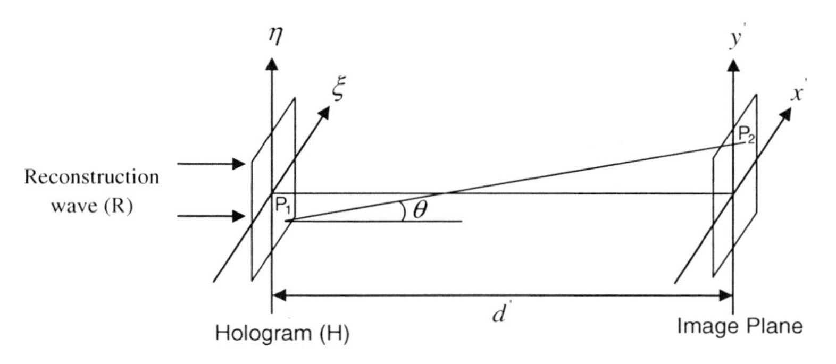

Figure 2.3 Diffraction geometry of a hologram reconstruction

reconstructed wavefield $U\left(n'_x, n'_y\right)$, at the image plane (x', y'), at distance d' from the hologram plane, is given by the Fresnel diffraction equation as shown in [4]:

$$U(n'_x, n'_y) = \frac{e^{ikd'}}{i\lambda d'} \int_{-\infty}^{\infty} \int_{-\infty}^{\infty} H(n_x, n_y) R(n_x, n_y) \exp\left[\frac{i\pi}{\lambda d'}\left\{(x' - \xi)^2 + (y' - \eta)^2\right\}\right] d\xi d\eta \tag{2.4}$$

where $k = \frac{2\pi}{\lambda}$, is the wave number.

The Fresnel Transform Method

The impulse response $g(n_x, n_y)$ of the coherent optical system can be defined as

$$g(n_x, n_y) = \frac{e^{ikd'}}{i\lambda d'} \exp\left\{\frac{i\pi}{\lambda d'}(\xi^2 + \eta^2)\right\} \tag{2.5}$$

Equations 2.4 and 2.5 can also be written as:

$$U(n'_x, n'_y) = e^{i\pi d'(\nu^2 + \mu^2)} \int_{-\infty}^{\infty} \int_{-\infty}^{\infty} H(n_x, n_y) R(n_x, n_y) g(n_x, n_y) \exp[-2\pi i\{\xi\nu + \eta\mu\}] d\xi d\eta \tag{2.6}$$

Here $\nu = \frac{x'}{\lambda d'}$ and $\mu = \frac{y'}{\lambda d'}$ are the spatial frequencies present in the hologram. Thus the reconstructed field is simply the Fourier transform of the product of the hologram, the reconstruction wave and the impulse response function, that is,

$$U(n'_x, n'_y) = \Im\{H(n_x, n_y)R(n_x, n_y)g(n_x, n_y)\} \tag{2.7}$$

If the hologram recorded by the CCD contains $N_x \times N_y$ pixels with pixel size Δ_x and Δ_y along the coordinates respectively, then the reconstructed field defined by Equation 2.7 is converted to finite sums as shown in [5]:

$$\begin{aligned} U\left(n'_x, n'_y\right) = &\frac{e^{ikd'}}{i\lambda d'} e^{i\pi\lambda d'\left(\frac{k^2}{N_x^2\Delta_x^2}+\frac{l^2}{N_y^2\Delta_y^2}\right)} \\ &\times \sum_{n_x=0}^{N_x-1}\sum_{n_y=0}^{N_y-1} H\left(n_x, n_y\right) R\left(n_x, n_y\right) e^{\left[\frac{i\pi}{\lambda d'}\left(n_x^2\Delta_x^2+n_y^2\Delta_y^2\right)\right]} e^{\left[-2\pi i\left(\frac{n_x k}{N_x}+\frac{n_y l}{N_y}\right)\right]} \end{aligned} \tag{2.8}$$

where $k = 0, 1, \ldots, N_x - 1$ and $l = 0, 1, 2, \ldots, N_y - 1$. Equation 2.8 is the discrete Fresnel transformation. The matrix $U(n'_x, n'_y)$ is the discrete Fourier transform of the product of $H(n_x, n_y)$, $R(n_x, n_y)$ and $\exp\{(i\pi/\lambda d')({n_x}^2{\Delta_x}^2 + {n_y}^2{\Delta_y}^2)$. Thus, the calculation of the reconstructed wave field can be done effectively by using the fast Fourier transform (FFT) algorithm. The pixel size of the numerically reconstructed image varies with the reconstruction distance and is given by:

$$\Delta'_x = \frac{\lambda d'}{N_x \Delta_x}, \quad \Delta'_y = \frac{\lambda d'}{N_y \Delta_y} \tag{2.9}$$

The Convolution Method

Reconstruction by the convolution approach is useful if the pitch of the reconstructed image has to be independent of the reconstruction distance. It can be shown that the diffraction integral (Equation 2.6) becomes a convolution for a linear space invariant system. Then the numerically reconstructed wave field can be written as:

$$U\left(n'_x, n'_y\right) = [H(n_x, n_y)R(n_x, n_y)] \otimes [g(n_x, n_y)]$$

here, $\otimes$ indicates a two-dimensional convolution. The reconstructed wavefield can be calculated by using the convolution theorem:

$$U(n'_x, n''_y) = \Im^{-1}[\Im\{H(n_x, n_y)R(n_x, n_y)\}\Im\{g(n_x, n_y)\}] \tag{2.10}$$

The calculation of the Fourier transforms can be performed effectively by the FFT algorithm. The Fourier transform of the impulse response is called the transfer function $G(n_x, n_y)$ of the free space propagation and is numerically defined as shown in [5]:

$$G(n_x, n_y) = \exp\left\{ \frac{2\pi i d'}{\lambda} \sqrt{1 - \frac{\lambda^2(n_x + N_x^2\Delta_x^2/2d'\lambda)^2}{N_x^2\Delta_x^2} - \frac{\lambda^2(n_y + N_y^2\Delta_y^2/2d'\lambda)^2}{N_y^2\Delta_y^2}} \right\} \tag{2.11}$$

Then the reconstructed wave field is given by:

$$U(n'_x, n''_y) = \Im^{-1}[\Im\{H(n_x, n_y)R(n_x, n_y)\}G(n_x, n_y)\}] \tag{2.12}$$

The pixel size of the reconstructed image by convolution method is the same as the pixel size of the CCD and does not vary with the reconstruction distance, that is, $\Delta'_x = \Delta_x$ and $\Delta'_y = \Delta_y$.

The Fresnel approach to reconstruction (Equation 2.7) is fast and useful when the object is larger than the CCD, but variation of the pixel size with the reconstruction distance creates problems in application, such as particles sizing and color holography. The pixel size of the reconstructed image remains constant with the reconstruction distance in the convolution approach but it is used when the object size is smaller than the CCD. In the case of a large object, zero padding of holograms is necessary before implementing the reconstruction by the convolution method.

The image intensity I and phase ϕ on the real image plane can be calculated as:

$$I = |U(n_{x'}, n_{y'})|^2 \quad \text{and} \quad \phi = \arctan\left[\mathrm{Im}(U(n_{x'}, n_{y'}))/\mathrm{Re}(U(n_{x'}, n_{y'}))\right] \tag{2.13}$$

2.2 Reflection Digital Holographic Microscope (DHM) Systems Development

2.2.1 Optical Systems and Methodology

2.2.1.1 In-Line Based DHM System

The optical system arrangement of the in-line digital holographic microscopy system is shown in Figure 2.4. A laser beam is divided into two parts and the intensities of the beams are adjusted to maximize the contrast of the interference fringes. A long distance microscope (LDM) is used in the object beam

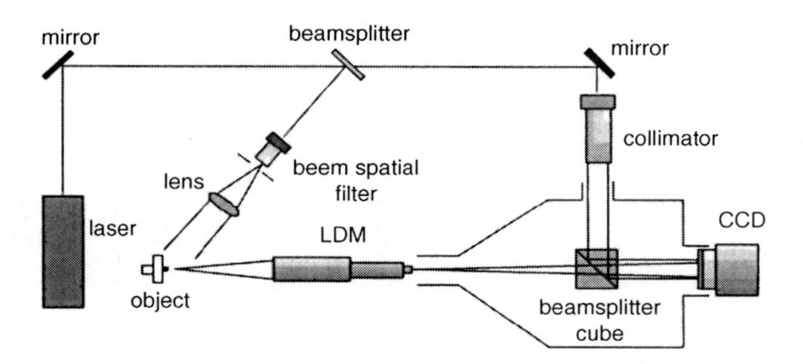

Figure 2.4 In-line digital holographic microscope set-up (*Source*: (6) Reproduced by permission of © 2001 Optical Society of America)

path to magnify the object's micro-features, as shown in [6]. The magnified object beam interferes with the in-line reference wave by using a cube beamsplitter and then is recorded numerically using a CCD sensor, and is stored in an image-processing system.

The numerically reconstructed wave field, because of in-line geometry, represents the real-image wave, the twin-image wave and the zero-order term along the optical axis and can be written as:

$$U(n_{x'}, n_{y'}) = U_{real-image-wave}(n_{x'}, n_{y'}) + U_{twin-image-wave}(n_{x'}, n_{y'}) + U_{zero-order-wave}(n_{x'}, n_{y'}) \tag{2.14}$$

The image intensity I and phase ϕ on the real image plane can be calculated as defined in Equation 2.13. But, due to the simultaneous presence of the twin-image wave and the zero-order term, the intensity and phase of the real image cannot be used directly for imaging or measurement applications. The double exposure method can be used for the imaging and measurement applications for the in-line digital holography system. Here one hologram is recorded, corresponding to the reference state of the object, and the second hologram is recorded, corresponding to the measurement state. The double exposure methodology is summarized in Figure 2.5.

In digital holographic interferometry, for static and dynamic measurement, both the amplitude as well as the phase difference of the two holograms recorded at two different states provide the characteristics of the dynamic phenomenon. The instantaneous dynamic deformation of any object can be measured using a high speed CCD camera, in this way, digital holograms are recorded corresponding to the different deformations of the objects and their subtraction provides the dynamic changes. When an object placed on the plane

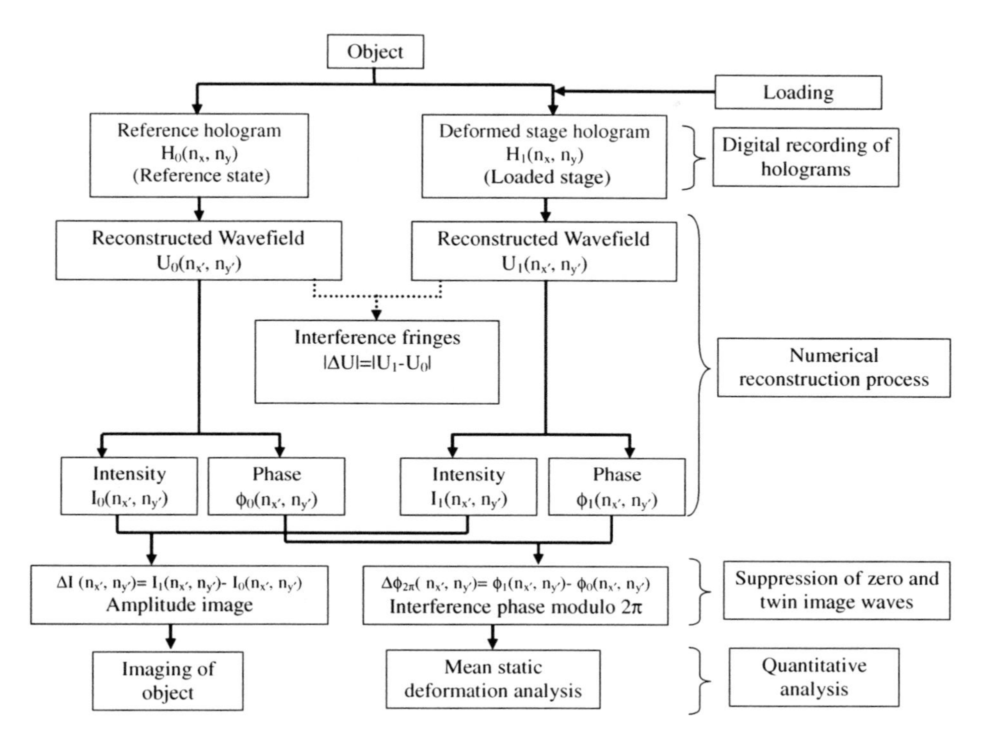

Figure 2.5 The double exposure method of in-line digital hologram for imaging and measurements

(x, y) is illuminated by the coherent beam, the light wave reflected, scattered or diffracted from the object surface can be written as:

$$O(n_x, n_y) = O_0(n_x, n_y)e^{i\phi_0(n_x,n_y)} \tag{2.15}$$

where $O_0(n_x, n_y)$ is the complex amplitude of the light and $\phi_0(n_x, n_y)$ is the phase representing the object surface properties. This object wave interferes with the in-line reference wave and a hologram is recorded using CCD. For different deformations of the object, the amplitude of the object wave is the same, however, the phase is changed for the corresponding deformation.

During reconstruction, the zero-order wave and the twin image overlap with this real image wave. Apart from a magnification term, the reconstructed wave can be written as:

$$U(n_{x'}, n_{y'}) = O_0(n_{x'}, n_{y'})e^{i\phi(n_{x'},n_{y'})} + U_{I+II} \tag{2.16}$$

where the first part of the right-hand side is the reconstructed real image wave and U_{I+II} is the sum of the background reference wave (zero-order wave) and the twin image wave. The background noise in the final reconstructed real image of the object is due to the zero-order wave and the out-of-focus twin image wave, and does not change with the different deformation states of the same object. Thus, in order to eliminate the background reference beam and the twin image, phases of the reconstructed image wave are calculated corresponding to the two different deformation states of the object and then subtracted. The numerically subtracted phase becomes the modulo 2π interference phase, which provides the deformation map. If $\phi_0(n_x, n_y)$ is the phase corresponding to the static state of the object (the reference state) and $\phi_1(n_x, n_y)$ is the phase corresponding to the deformation state, then the subtraction of the phases of numerically reconstructed in-line holograms can be written as:

$$\Delta\phi_{2\pi} = \phi_1 - \phi_0 \tag{2.17}$$

Vibration Analysis by the Time-Averaged Method

For a sinusoidally vibrating object (Figure 2.6) in the (x, y) plane, the instantaneous object wave $O'(n_x, n_y, t)$ at any instant scattered from the vibrating object is:

$$O'(n_x, n_y, t) = O_0(n_x, n_y)e^{i\phi_0(n_x,n_y)}e^{i[\vec{K}\cdot\vec{z}\,(n_x,n_y,t)]} \tag{2.18}$$

where $\phi_0(n_x, n_y)$ is the phase representing the mean deformation state of the vibrating object, $\vec{K}$ is the sensitivity vector and $\vec{z}\,(n_x, n_y)$ is the amplitude of vibration.

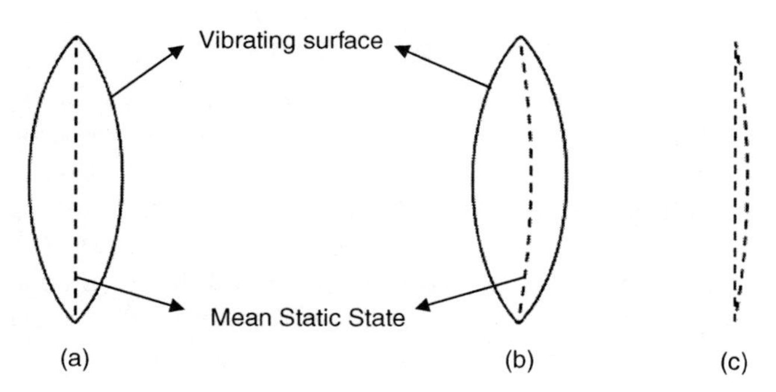

Figure 2.6 (a) Sinusoidally vibrating object; (b) Vibrating object with a change in the mean static state (c) Change in the mean static state

For a time-averaged recording, the frame capture time τ of the CCD should be longer than the period of object vibration. The time-averaged object wave thus becomes:

$$O(n_x, n_y) = \frac{1}{\tau}\int_0^{\tau} O'(n_x, n_y, t)d\tau \quad (2.19)$$

$$O(n_x, n_y) = O_0(n_x, n_y)\exp\{i\phi(n_x, n_y)\} \times J_0\{\vec{K} \cdot \vec{z}_v(n_x, n_y)\} \quad (2.20)$$

where J_0 is the zero-order Bessel function and $\phi(n_x, n_y)$ represents the phase of the object wave which contains the information both about the mean static deformation and the zeros of the Bessel function and is defined as:

$$\phi(n_x, n_y) = \phi_o(n_x, n_y) + \phi_J(n_x, n_y) \quad (2.21)$$

where $\phi_0(n_x, n_y)$ contains object surface data and $\phi_J(n_x, n_y)$ is the time-averaged phase. The in-line reference wave interferes with the object wave from the vibrating object and thus the time-averaged hologram is recorded by the CCD.

The reconstructed wave, in this case, can be written as:

$$U(n_{x'}, n_{y'}) = O_0(n_{x'}, n_{y'})e^{i\phi(n_{x'}, n_{y'})}J_0[\vec{K} \cdot \vec{z}\,(n_{x'}, n_{y'})] + U_{I+II} \quad (2.22)$$

Here the first term on the right-hand side is the reconstructed real image wave and U_{I+II} is the background noise. This background noise in the final reconstructed real image of the object is due to the zero-order wave and the out-of-focus twin image wave, and it does not change significantly with different deformation states of the same object. However, since the speckle pattern changes stochastically between exposures, the background noise may not be exactly equal.

The amplitude and phase of the numerically reconstructed real image wave are, apart from a magnification term, as follows:

$$A(n_{x'}, n_{y'}) = |U(n_{x'}, n_{y'})| = O_0(n_{x'}, n_{y'})\, J_0 \,\{\vec{K} \cdot \vec{z}_v(n_{x'}, n_{y'})\} \quad (2.23)$$

and

$$\phi\,(n_{x'}, n_{y'}) = \arctan\frac{\mathrm{Im}\,(U(n_{x'}, n_{y'})}{\mathrm{Re}\,(U(n_{x'}, n_{y'})} = \phi_o(n_{x'}, n_{y'}) + \phi_J(n_{x'}, n_{y'}) \quad (2.24)$$

In order to suppress the background noise, two time-averaged in-line holograms of the object are recorded, one corresponding to the reference state (either static or vibration) and other is the vibration state. Consider an object vibrating at an angular at two different amplitudes $\vec{z}_1$ and with $\vec{z}_2$ corresponding to the two states. The phases of the mean deformation states are ϕ_1 and ϕ_2 respectively. On numerical reconstruction, the wave fields can be written as:

Reference state:

$$U_1(n_{x'}, n_{y'}) = O_0(n_{x'}, n_{y'})e^{i\phi_1(n_{x'}n_{y'})}J_0[\vec{K} \cdot \vec{z}_1(n_{x'}, n_{y'})] + (U_{I+II})_1 \tag{2.25a}$$

Vibration state:

$$U_2(n_{x'}, n_{y'}) = O_0(n_{x'}, n_{y'})e^{i\phi_2(n_{x'},n_{y'})}J_0[\vec{K} \cdot \vec{z}_2(n_{x'}, n_{y'})] + (U_{I+II})_2 \tag{2.25b}$$

$(U_{I+II})_1$ and $(U_{I+II})_2$ do not change for different deformations of the same object. Thus subtraction of wave field provides the information about the vibration behavior of the objects free from background noise. To do this, the amplitude and phase of the individual wave fields are extracted first. The difference in amplitudes then provides the Bessel type vibration fringes while the difference of phase gives the mean deformation, as shown in [7].

2.2.1.2 Off-Axis Based DHM Systems

The schematic of optical geometry of reflection digital holographic microscope is shown in Figure 2.7. A single mode fiber is coupled to a laser source. The

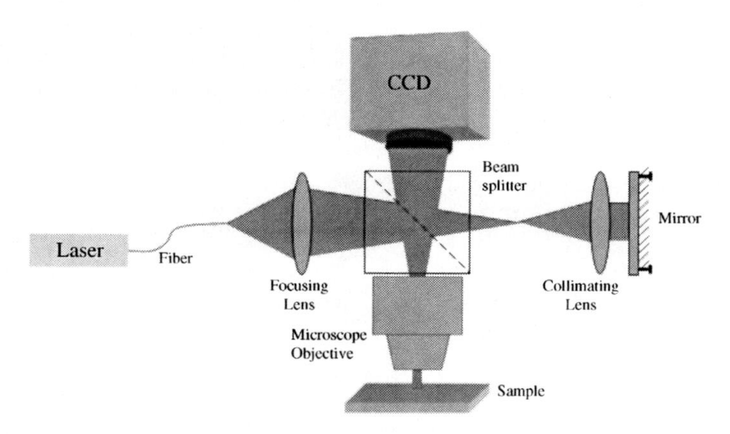

Figure 2.7 The off-axis reflection digital holography microscopy system (*Source*: (8) Reproduced by permission of © 2009 Optical Society of America)

laser beam coming from the fiber is focused, using focusing lens and is split into two parts by using the beam splitter (BS). The microscopic objective (MO) is placed on one side of the BS and the focusing lens is adjusted so that the object beam coming from the MO is collimated. The advantage is that it provides the same curvature of the object wave for different positions of the object from the MO. The other beam also gets collimated and is reflected by the mirror. The tilt screws of the mirror control the angle of reference wave (reflected by the mirror). The object beam and reference beam, after reflection, interfere (this is called a hologram) and are recorded by the CCD. The advantage of the geometry is the same wavefront of the object and reference waves and thus the spherical aberration of the MO are automatically compensated.

Pre-Processing of Digital Off-Axis Hologram

Off-axis geometry is suitable for separating the zero-order term and the twin image terms from the real image wave by giving an offset angle between the object and the reference waves. Before the reconstruction of the digital off-axis hologram, the pre-processing procedure can be performed to reconstruct only the real image wave. The pre-processing method is shown in Figure 2.8. Basically the +1 order from the frequency spectrum of the off-axis hologram is selected, which represents the information about the real image wave. The

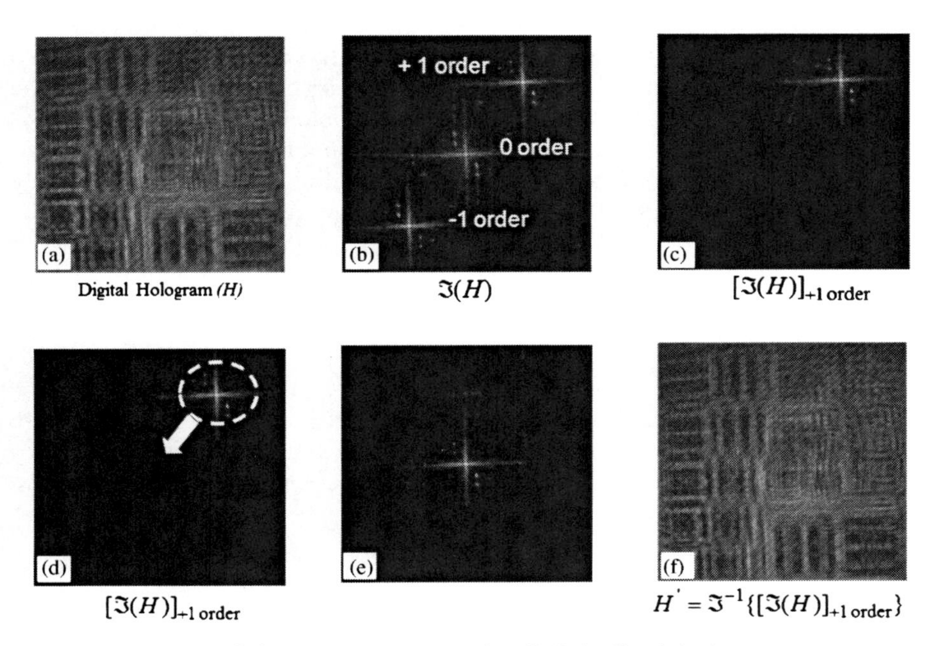

Figure 2.8 Pre-processing of a digital off-axis hologram

inverse Fourier transform of the centered first order is called the pre-processed hologram, which is free from zero-order and twin image waves.

3D Profile Measurement

Numerically reconstructed phase data is used to measure the 3D profile of the MEMS samples. The sample height/depth t can be written as:

$$t = \frac{\lambda\phi}{4\pi} \tag{2.26}$$

The direct phase value can be converted into height/depth values if the measurement is less than half the wavelength of the source used. For larger measurement values, the phase unwrapping method is used to remove the phase jumps. In the case of steep height measurements of more than half the wavelength, the phase jumps cannot be identified accurately and this creates an error in the measurement results.

Deflection/Deformation Measurement

Deflection/deformation measurement can be performed using the digital holography interferometry method. In it, two holograms are recorded, corresponding to the two different deflection/deformation states of the sample. The reconstructed phases of the two corresponding states are calculated and their subtraction provides the deflection/deformation that occurs between the two states. The measurement speed between the two exposures is decided by the frame rate of the CCD sensor used to record the hologram. Thus, the fast dynamic deformation of sample can be measured using a high speed CCD camera. When the sample surface is illuminated by the beam, the numerically reconstructed object wave corresponding to the two deformation states can be written as:

$$O_0(n_{x'}, n_{y'}) = O_0(n_{x'}, n_{y'})e^{i\phi_0(n_{x'}, n_{y'})} \tag{2.27}$$

$$O_1(n_{x\prime}, n_{y\prime}) = O_0(n_{x\prime}, n_{y\prime})e^{i\phi_1(n_{x\prime}, n_{y\prime})} \tag{2.28}$$

The numerical reconstructed phase subtraction of the reference state from the deformed state represents the modulo 2π interference phase, which provides the deformation map. The subtraction of the phases provides the interference pattern written as:

$$\Delta\phi_{2\pi} = \phi_1 - \phi_0 \tag{2.29}$$

After phase unwrapping and using Equation 2.29 the exact amount of deformation can be calculated.

Vibration Measurements

On a numerical reconstruction of a time-averaged hologram using the off-axis digital holography system, the reconstructed real image wave is written as:

$$U(n_{x'}, n_{y'}) = O_0(n_{x'}, n_{y'})e^{i\phi(n_{x'},n_{y'})}J_0[\vec{K} \cdot \vec{z}\,(n_{x'}, n_{y'})] \tag{2.30}$$

The amplitude of the numerically reconstructed real image wave is written as follows:

$$A(n_{x'}, n_{y'}) = |U(n_{x'}, n_{y'})| = O_0(n_{x'}, n_{y'})J_0\{\vec{K} \cdot \vec{z}_v(n_{x'}, n_{y'})\} \tag{2.31}$$

and the phase is:

$$\phi(n_{x'}, n_{y'}) = \arctan\frac{\mathrm{Im}\,(U(n_{x'}, n_{y'})}{\mathrm{Re}\,(U(n_{x'}, n_{y'})} = \phi_o(n_{x'}, n_{y'}) + \phi_J(n_{x'}, n_{y'}) \tag{2.32}$$

Here amplitude is modulated by the zero-order Bessel function which provides the data on the mode shape and amplitude of vibrations of the object. The numerically reconstructed phase from the time-average holograms is a combination of the phase due to the mean static state data ϕ_o, and the time-average phase ϕ_J. ϕ_o varies from $-\pi$ to $+\pi$, whereas ϕ_J is the binary phase (with values 0 and $\pm\pi$) that changes with the zeros of the Bessel function. However, in the presence of both static deformation and vibrations, the phase of the time-averaged hologram represents the mixing of the mean deformation and the time-averaged fringes represented by the reconstructed phase information.

2.2.1.3 Lens-Less DHM Systems

The schematic of a lens-less digital holographic reflection microscope is shown in Figure 2.9 [9]. A point source provides a diverging beam, which is centered about the optical axis. The divergence of the beam provides the lens-less magnification in the system. A beam splitter is used to separate the diverging beam, reflecting a portion towards the test object and transmitting a portion to a mirror located behind the beam splitter. The scattered light from the test object is called the object beam and the reflection beam from the mirror is the reference beam. The beam splitter transmits the object beam towards, and reflects the reference beam to, a CCD sensor. The interference pattern between the object and the reference beam is converted to a digital signal by the CCD, called a digital hologram, which is then numerically processed to determine the desired characteristics of the object. The interference angle can be controlled by tilting the reference mirror, and this can be used in the system in

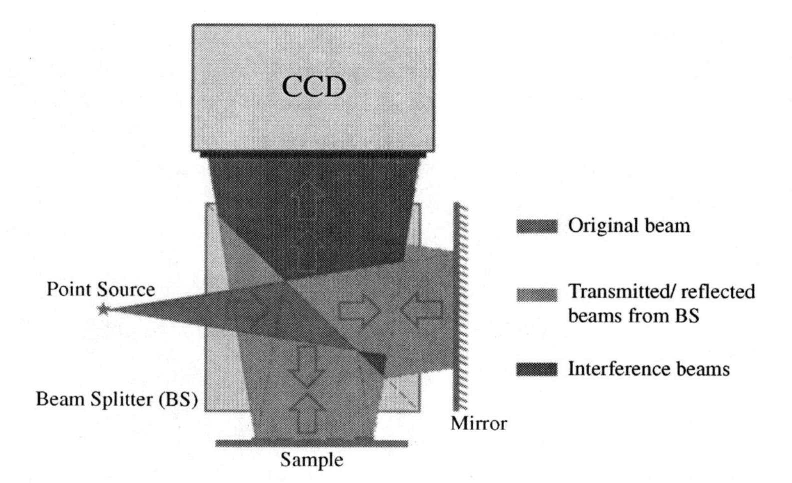

Figure 2.9 Schematic of the lens-less digital holographic microscope (*Source*: (9,10) Reproduced by permission of © 2010 Society of Photo-optical Instrumentation Engineers)

both in-line and off-axis mode. Off-axis geometry is particularly useful for phase reconstruction from a single hologram. This system presents lens-less reflection microscopy geometry by using a very simple and compact optical set-up best suited for the study of micro-size samples. The novelty of the system is the compact optical geometry with the minimum number of optical components and because it is suitable to be developed as a handheld system. The other advantage of the system is because of the same wavefront of the object and the reference beams, and because it does not provide any spherical phase aberrations during the reconstruction of the hologram.

The system presented here is most suitable for the study of highly specular objects and provides the 3D measurement from the direct phase calculation by reconstructing the single off-axis hologram. The ratio of the distances from the point source to the CCD and from the point source to the sample defines the geometrical magnification of the system. Since the distance from the point source to the object is constrained by the beam splitter, so it restricts the magnification of the system. The system is not perfectly suited to the analysis of the diffused surface samples, because the object beam is scattered from the sample surface and its amplitude is significantly smaller than the reference beam amplitude and thus does not provide a good contrast of the interference pattern when recorded by the CCD.

This optical system design is housed in a compact casing as shown in Figure 2.10. The illumination source is a single mode fiber end which provides the diverging laser beam attached to the casing. The other end of the fiber is

Figure 2.10 Packaged handheld digital holographic microscopy system (*Source*: (11) Reproduced by permission of © 2010 Society of Photo-optical Instrumentation Engineers)

attached to the laser. The wavelength of the light is chosen according to the application or the required lateral resolution. For the system shown here, a fiber-coupled laser diode with the wavelength 642 nm is used. The diverging beam is delivered from the end of an optical fiber, where the smaller the diameter of the fiber (for example, in the order of wavelength), the greater the cone of the emitted light coming through it. Therefore a smaller diameter fiber may increase the numerical aperture (NA) of the system and thus the system resolution.

2.3 3D Imaging, Static and Dynamic Measurements

2.3.1 Numerical Phase and 3D Measurements

Numerically reconstructed phase data is used to measure the 3D profile of the objects. The object height/depth t can be written as:

$$t = \frac{\lambda\phi}{4\pi} \tag{2.33}$$

The direct phase value can be converted into height/depth values if it measures less than half the wavelength of the source used. For larger measurement

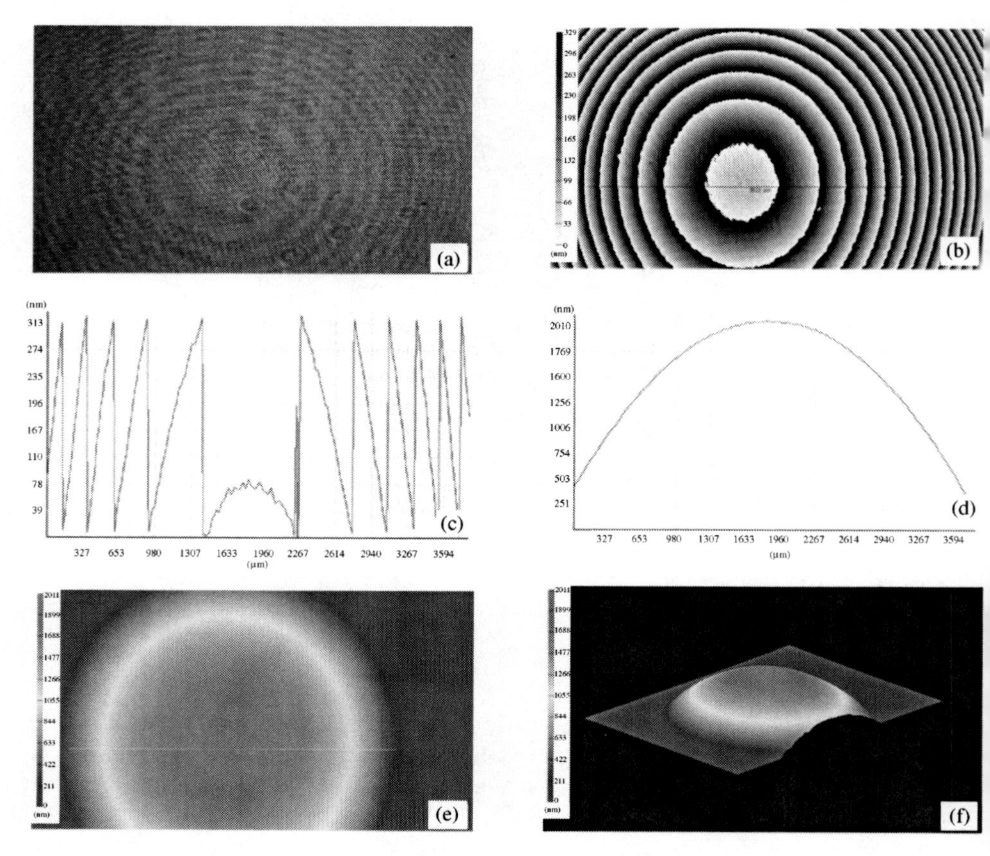

Figure 2.11 Semiconductor wafer warpage. (a) Digital hologram; (b) Modulo-2π phase image; (c) Wrapped phase profile; (d) Unwrapped phase image; (e) Surface profile; (f) 3D perspective (*Source*: (12) Reproduced by permission of © 2008 Society of Photo-optical Instrumentation Engineers)

values, the phase unwrapping method is used to remove the phase jumps. In the case of steep height measurements of more than half the wavelength, the phase jumps cannot be identified accurately and this creates an error in the measurement results.

In Figure 2.11, a semiconductor device whose surface shows warpage and is detached from the adhesive bond was inspected. The recorded hologram is shown in Figure 2.11a, the reconstructed phase image is shown in Figure 2.11b which shows the modulo -2π phase which can be seen clearly in Figure 2.11c. To reconstruct the 3D map from the unwrap phase, the phase unwrapping is done and shown in Figures 2.11d–e. The warpage revealed by the unwrapped phase image shown in the 3D representation in shown in Figure 2.11f.

2.3.2 Digital Holographic Interferometry

2.3.2.1 Static Deflection/Deformation Measurement

Here deflection analysis of a piezo-actuated circular micro-membrane diaphragm of size 1 mm is studied for different applied voltages. The diaphragm is excited by applying a DC driving voltage across the piezoelectric layers. The electrodes were connected to the power supply and holograms were recorded corresponding to the different applied voltages between the electrodes. The phase data is reconstructed by the recorded hologram at different voltages. The subtraction of the phases of the deformed state (corresponding to the applied voltage) and the reference state (without applying voltage) provide the deformation fringes. Figure 2.12 shows the deformation measurement for the

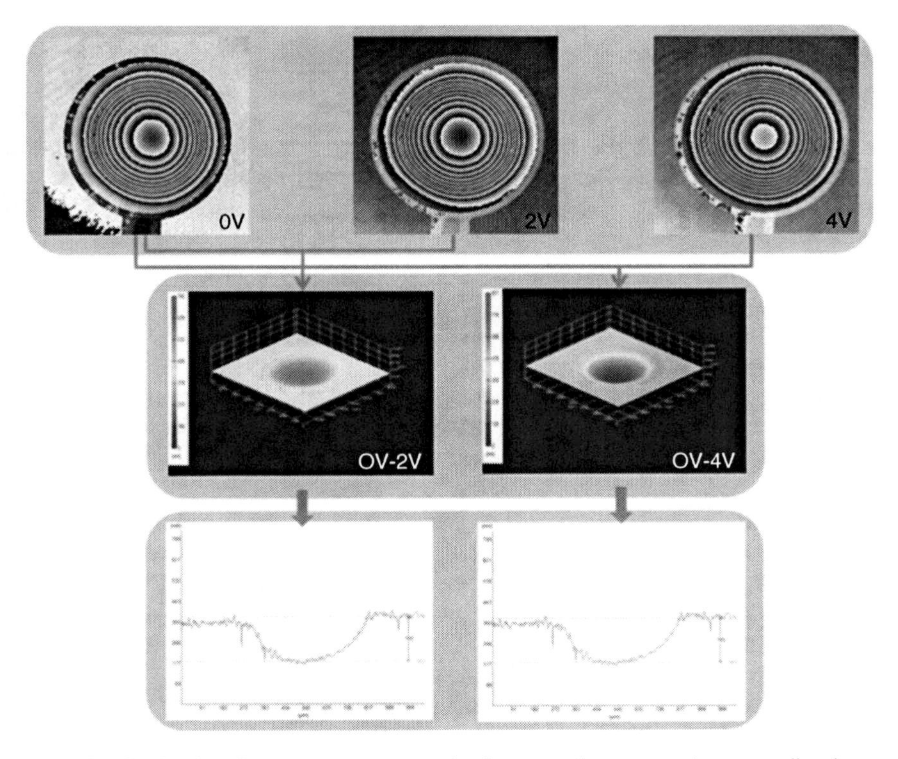

Figure 2.12 Deflection measurements in a micro-membrane diaphragm (*Source*: (10) Reproduced by permission of © 2010 Society of Photo-optical Instrumentation Engineers)

two different voltages. The phase images of the diaphragm corresponding to the 0 V, 2 V and 4 V are shown in the first row. To get the deformation map at the required voltage, the phase image obtained at that voltage is subtracted from the phase obtained at the reference voltage (0 V). The deformation 3D map of the diaphragm is shown in the second row, corresponding to 2 V and 4 V respectively, and the corresponding line profile is shown in the third row which shows the deformation values are 193 nm and 423 nm respectively.

2.3.2.2 Dynamic Measurement

Digital dynamic holographic interferometry is different from static digital holographic interferometry in that both the amplitude as well as the phase difference of the two holograms recorded at two different states provide the imaging characteristics of the dynamic phenomenon. For vibration measurements, a time-averaged digital hologram is used to record the dynamic behavior of periodically vibrating objects. This is not suitable for a transitional phenomenon wherein traditional double exposure holographic interferometry can be adopted using a high-speed camera.

Dynamic digital holography has also attracted great interest in recent years. There are two approaches to this – the first is to use a pulse laser or high speed camera to record multiple frames which are then processed much like the static case. Methods such as pulse or stroboscopic digital holography require precise synchronization of the light source, the specimen and the recording device, which makes the system complex. The second and more interesting approach is to use the time-average method, which does not require any high speed camera or a pulsed laser.

Vibration Analysis

An experimental study of vibration analysis of an aluminum membrane object is presented here based on the methodology presented in Section 2.4 for in-line digital holography. The experimental set-up for in-line digital holography used for the vibration study is shown in Figure 2.13a. A frequency doubled Nd-YAG laser operating at 0.532 μm is split into two beams by a variable beam splitter. The object beam is expanded and illuminates the object, while the other is the collimated reference beam. The object, a thin aluminum membrane bonded to a circular metal ring of outer and inner diameters 13 mm and 10 mm respectively (Figure 2.13b), is attached to an earphone.

The earphone is excited, using a frequency generator which can be tuned to vibrate at different sinusoidal frequencies and amplitudes. The static deformation of the membrane is controlled by varying the amplitude of the excitation signal by using the frequency generator. An 8-bit digital CCD with 2029 × 2044 square pixels 9 μm in size is used to record the holograms at 30 frames per second. The object is placed 500 mm from the CCD, which satisfies the

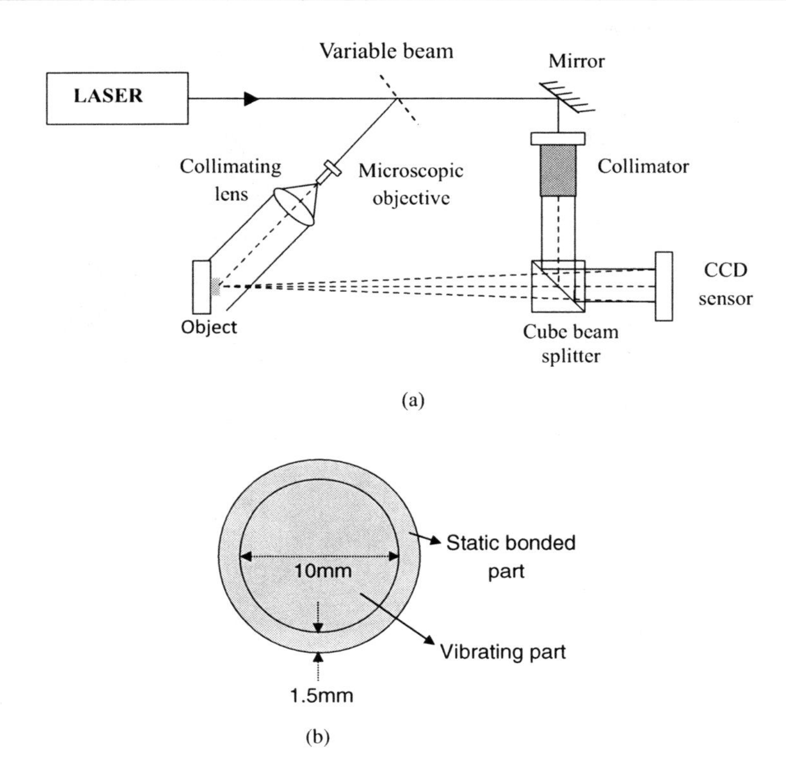

Figure 2.13 (a) In-line digital holographic set-up for vibration measurement; (b) Schematic of the vibrating membrane (*Source*: (13) Reproduced by permission of © 2006 Optical Society of America)

sampling theorem requirement for efficient use of the full sensor area. The in-line time-average digital holograms are recorded at different amplitudes of vibration and at frequencies ranging from 1 KHz to 10 KHz. Thus, the exposure time is greater than the period of vibration, giving rise to a time-average recording.

For the time-average recording, the amplitude is modulated by the J_0 function and the phase represents data about the mean static state of the object. If there is any change in the mean static state of the vibrating object at different amplitudes, then as previously discussed, the subtracted wave fields (or intensity) of two time-average holograms would show a mixture of time-average and static deformation fringes (Equation 2.25a). Double exposure holograms of the membrane are recorded at the same frequency but at different amplitudes. The subtraction of the reconstructed wave fields corresponding to the resonant frequencies 2.0 KHz, 5.25 KHz, 6.0 KHz, and 7.75 KHz is shown in Figure 2.14.

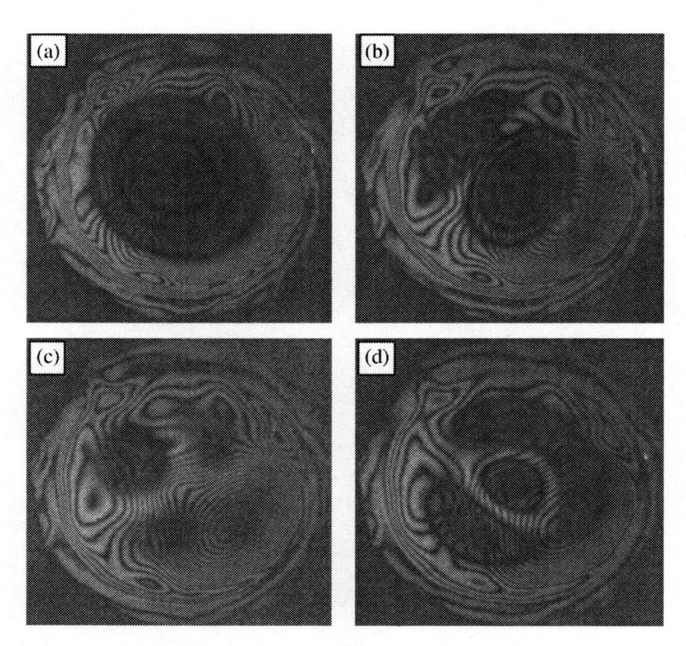

Figure 2.14 Subtracted wave fields of membrane vibrating at same frequency but different amplitude showing the mixing of vibration and mean static deformation fringes. Vibration frequencies at (a) 2.0 KHz (b) 5.25 KHz (c) 6.0 KHz (d) 7.75 KHz (*Source*: (13) Reproduced by permission of © 2006 Optical Society of America)

Two sets of fringes corresponding to Bessel fringes and the mean deformation fringes are clearly observed in Figure 2.14. The deformation fringes extend to the area where the diaphragm is bonded to the frame, indicating the bonding characteristics. The fringes in the clamped region are attributed to improper bonding of the diaphragm which may create the asymmetric mean deformations while applying vibrations. Furthermore, the fringe pattern appears to imply that the two sets of fringes are simply superposed. Indeed, if the two patterns were simply superposed, they could be readily separated by considering the amplitude and phase of the wave field subtraction.

To separate the Bessel fringes from the mean deformation fringes, it is thus necessary to first determine the amplitude and phase of the two time-average digital holograms and then subtract them, as given in Equation 2.25a. The mode patterns of the vibrating membrane corresponding to the same frequencies as discussed above are shown in Figure 2.15. These patterns are obtained by subtraction of reconstructed intensities from two time-average holograms recorded at the same vibration frequencies but with slightly different

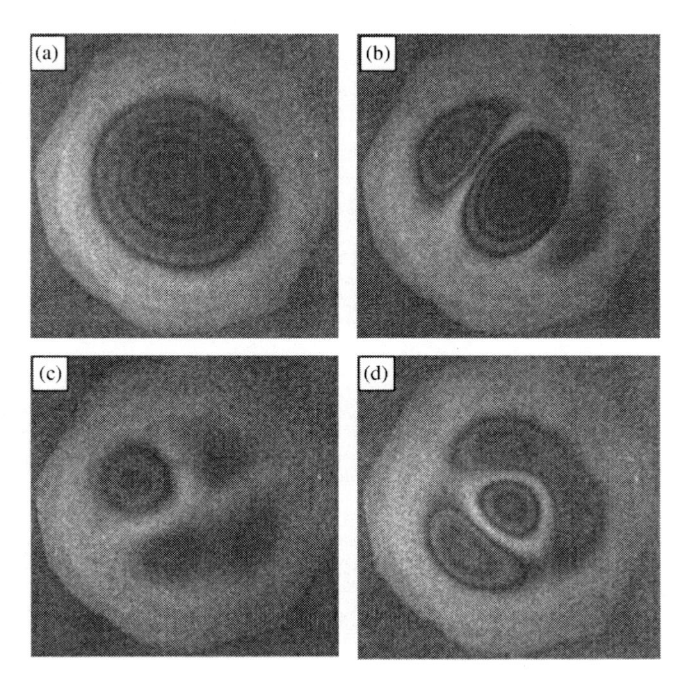

Figure 2.15 Vibration mode pattern at frequencies (a) 2.0 KHz; (b) 5.25 KHz; (c) 6.0 KHz; (d) 7.75 KHz (*Source*: (13) Reproduced by permission of © 2006 Optical Society of America)

amplitudes. Similar patterns can also be obtained by displaying the intensity of a single time-average hologram. These patterns are similar to those observed in conventional time-average holographic interferometry. For a single time-average hologram, the overlapping zero-order wave and twin image effect need to be eliminated by pre-processing the hologram. For the two exposure subtraction process, no such pre-processing is required.

The mean static state of the membrane under vibration changes the phase of the wave field. Hence to display only the difference in mean static state between two wave fields, the phase is extracted from each wave field and then subtracted as per Equation 2.25a. Figures 2.16a–d show the resulting phase difference representing only the difference in mean static state for the membrane vibrating at different frequencies. The difference in amplitudes between the two subtracted images for the four different frequencies is similar and hence the difference in mean static state fringes is identical.

Closer inspection of the fringe pattern indicates what appear to be additional phase effects due to the fact that the membrane is vibrating at resonance. These additional phase changes mask the mean deformation fringes in these regions.

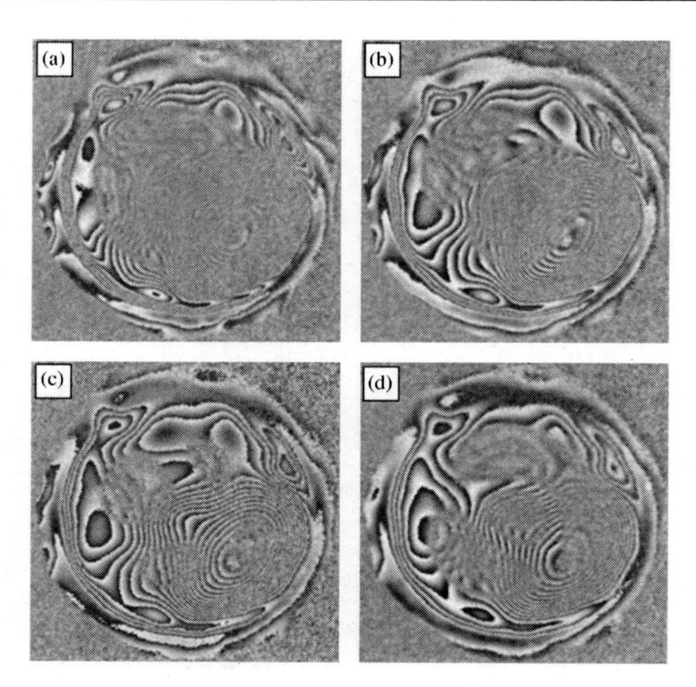

Figure 2.16 Subtraction of the phase gives the fringe pattern corresponding to mean static deformation during a change in vibration amplitudes (*Source*: (13) Reproduced by permission of © 2006 Optical Society of America)

Also, the amplitudes at resonant frequencies are much higher than for static deformation if the plate was not vibrating, as is to be expected. This is clear from Figure 2.17 which shows the same membrane, but only subject to static deformation at zero vibration frequency. Only the fringes due to difference in

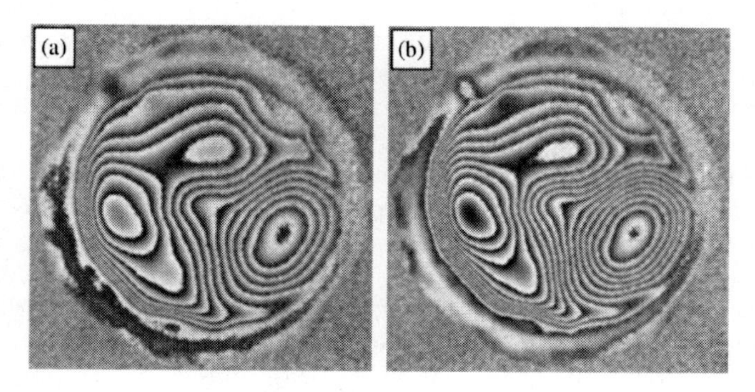

Figure 2.17 Mean static deformations at two different amplitudes of a stationary plate (*Source*: (13) Reproduced by permission of © 2006 Optical Society of America)

static states are clearly seen. Indeed, even if the plate were vibrating, as long as the amplitude of vibration is small, only fringes due to difference in the static state are observed. It is also noted that in this case (Figure 2.16), the difference in amplitude between the two exposures is two or three times larger than the amplitude difference for Figure 2.17.

2.4 MEMS/Microsystems Characterization Applications

New challenges for the measurement introduced by miniaturization of the test objects require the development of reliable advanced testing methods. Integration of the mechanical elements, electronics, sensors and actuators on a common silicon substrate by micromachining technology constitute a micro-electromechanical systems (MEMS). This has a wide range of applications in scientific and engineering fields. Characterization of the mechanical properties of MEMS structures at different stages of manufacturing is very significant. The purpose of MEMS testing is to provide feedback about device behavior, system parameters, and the material properties for the design and simulation process. Also dynamic testing is needed on the final devices to test their performance and characteristics. Characterization of the mechanical properties of MEMS structures is a challenging task.

The proposed digital holographic reflection microscope system is used as a novel instrumentation for the static and dynamic characterization of MEMS. The system aims to use optical full-field measurement methods together with combined computational and experimental methodologies, to implement comprehensive testability procedures in reliability testing of MEMS. It contributes to an understanding of MEMS mechanical and material issues during the whole component and system life.

2.4.1 3D Measurements

2.4.1.1 Surface Profile Measurements

3D profile measurements of MEMS circular micro-diaphragm are studied using the quantitative phase information obtained from a digital holographic microscope. The diaphragm was fabricated by bonding a piezoelectric plate onto an SOI (silicon on insulator) wafer with 20 μm thick device layer. The thickness of the piezoelectric layer was thinned down to about 40 μm by using chemical/mechanical polish. The silicon back was etched away by deep reactive ion etching. The hologram of the sample is recorded using the optical system as shown in Figure 2.1. The reconstructed amplitude contrast and phase contrast images of the three cantilevers are shown in Figures 2.18a and b respectively. The phase contrast image is significantly different from the

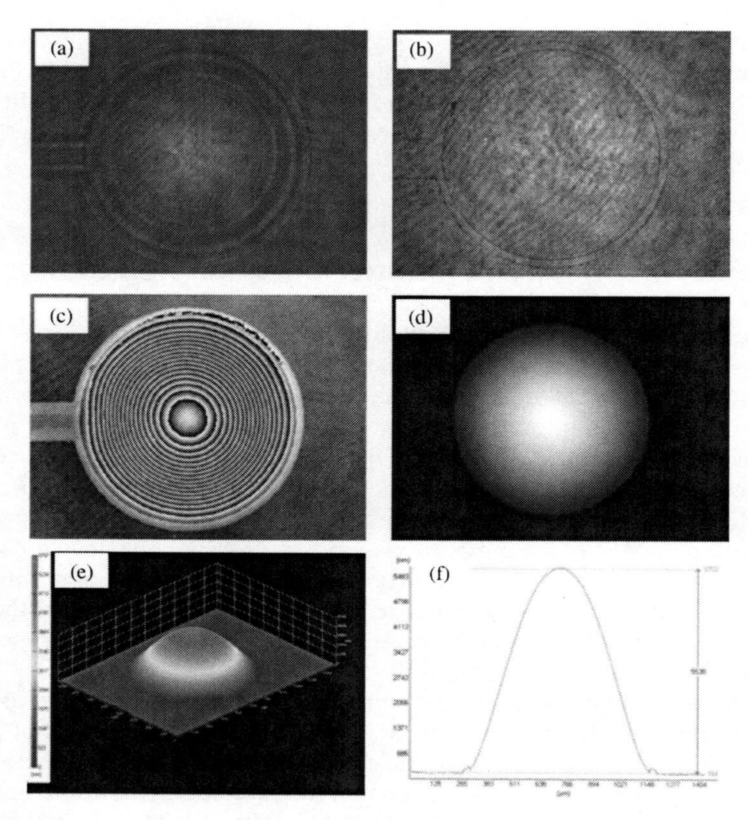

Figure 2.18 (a) Digital hologram; (b) Reconstructed amplitude image; (c) Reconstructed phase image; (d) Unwrapped phase image; (e) 3D profile of unwrapped phase; (d) Line profile (*Source*: [11] Reproduced by permission of © 2010 Society of Photo-optical Instrumentation Engineers)

amplitude image and used to measure the surface profile of the cantilevers. The wrapped phase can be seen clearly in Figure 2.18b because of the phase jumps of more than 2π. Figure 2.18c shows a 3D map of the unwrapped phase image, and the quantitative value of the phase was converted to a length scale. To calculate the deviation of the diaphragm, the line profile is plotted across the diameter of the unwrapped phase contrast image and is shown in Figure 2.18d. It is important to note that the amplitude, phase and 3D profile of the sample are obtained from the single hologram. The developed software provides near real-time reconstruction. Thus, this system is useful for the full field fast 3D measurement of sample surfaces.

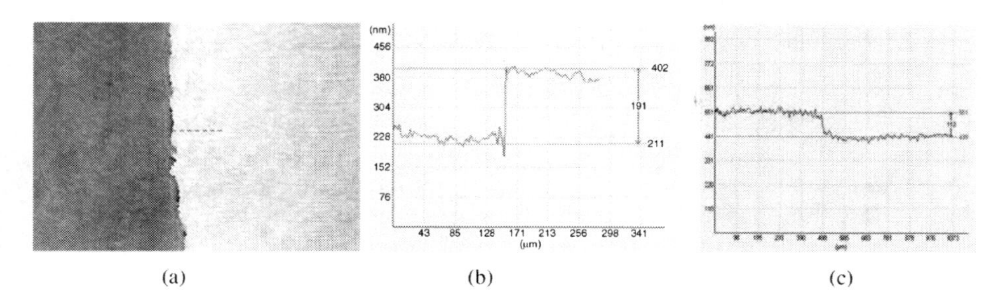

Figure 2.19 100 nm thin film thickness measurements (a) Amplitude image; (b) Phase image; (c) Line profile (*Source*: (8) Reproduced by permission of © 2009 Optical Society of America)

2.4.1.2 Deposition Height/Etching Profile Measurements

The first application of the system is shown for quantitative measurements of the deposition height of thin film. Two thin films of different thicknesses (100 nm and 200 nm) are deposited on a silicon wafer. Figure 2.19 shows the imaging and measurement results for the 100 nm film. Figure 2.19a shows the amplitude image which only shows the edges of the thin film (a sharp line), Figure 2.19b is the numerical phase contrast image which provides the quantitative measurements in 3D. The line profile shown in Figure 2.19c provides the height information from the phase contrast image. The phase contrast image and measurement results for 200 nm thickness film are shown in Figure 2.20.

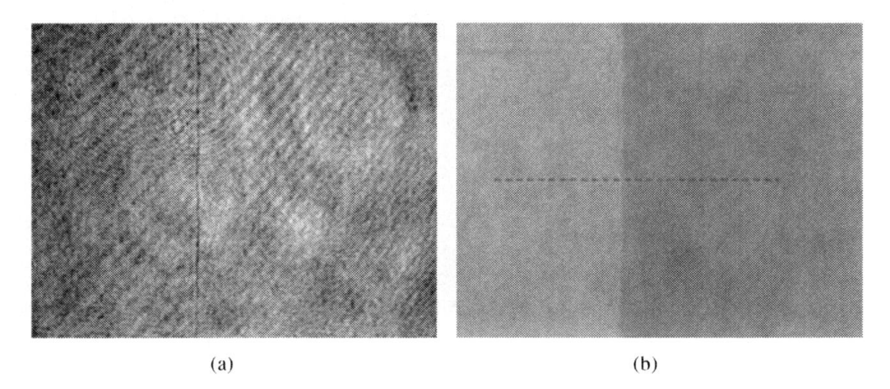

Figure 2.20 200 nm thin film thickness measurements (a) Phase image; (b) Line profile (*Source*: (8) Reproduced by permission of © 2009 Optical Society of America)

2.4.1.3 Accelerometer Device Inspection

The performance of a MEMS accelerometer when mounted on two different substrates, ceramic and PCB, is tested due to concerns about cracking at the interface of the device and the PCB substrate. The surface profile of the device after packaging is studied in both cases and shown in Figure 2.21. Figures 2.21a and b show the phase image and its 3D map of the device when mounted on a ceramic substrate and Figures 2.21c and d are for PCB substrate respectively. It can clearly be seen that the 3D surface map is flat for the ceramic substrate while that for the PCM substrate is warped, suggesting the introduction of mechanical stress at the interface between device and substrate. This study reveals the importance of the substrate on the characteristics of the MEMS device. Indeed, the warping of the device when mounted on a PCB substrate was a possible reason for the damage of this system as compared to the ceramic substrate.

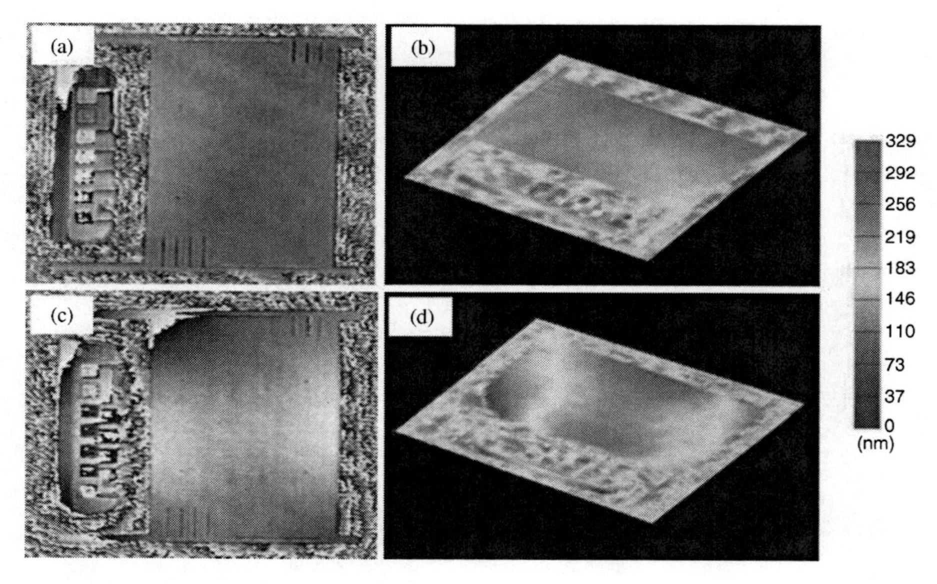

Figure 2.21 Accelerometer device analysis, (a) and (b) phase image and 3D profile of sensing area for ceramic substrate; and (c) and (d) phase image and 3D profile of sensing area for PCB substrate (*Source*: (8) Reproduced by permission of © 2009 Optical Society of America)

2.4.2 *Static Measurements and Dynamic Interferometric Measurement*

2.4.2.1 Deflection Measurements in Cantilevers

Aluminum nitride (AlN) films have piezoelectric properties, and represent an alternative to PZT films. Out-of-plane static deflection analysis of the AlN cantilevers (900 μm × 50 μm) was performed using the system presented here. The experimental set-up shown in Figure 2.9 is used for the deformation study of the cantilevers. The upper and lower electrodes were connected to the power supply and holograms were recorded corresponding to the different applied voltages between the electrodes. The phase data is reconstructed for the recorded hologram at different voltages. The subtraction of the phases, as defined in Equation 2.10, of the deformed state (corresponding to the applied voltage) and the reference state (without applying voltage) provides the deformation fringes. Figures 2.22a and c show the deformation fringes for the two

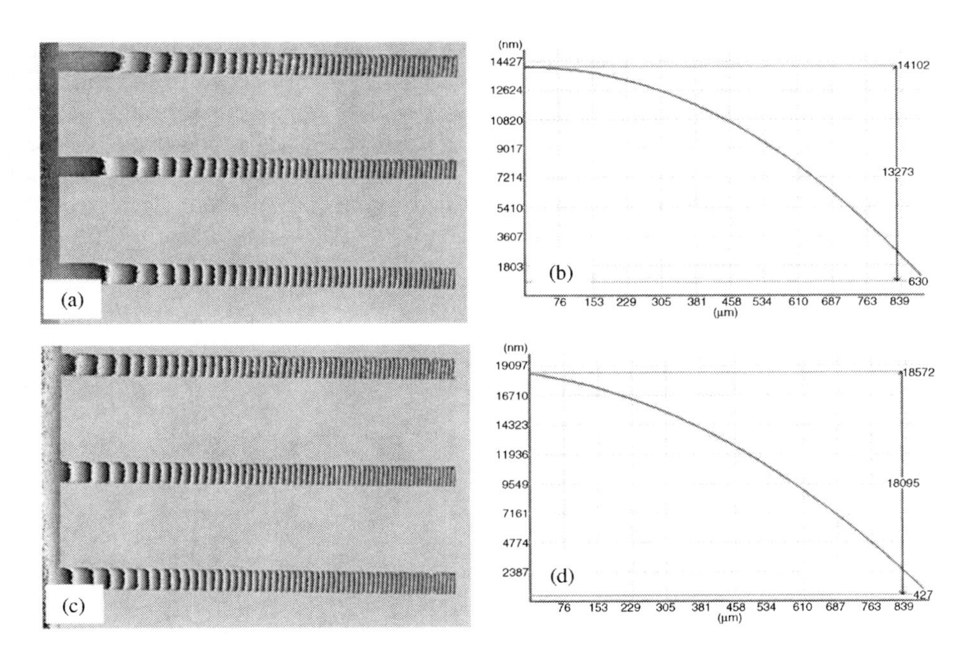

Figure 2.22 Static deflection measurements in cantilevers (a) and (c) Phase subtracted image from the reference state at different applied voltages, and deformation profile of center cantilevers are shown in (b) and (d) respectively (Reproduced by permission of © 2010 Society of Photo-optical Instrumentation Engineers)

different voltages. Deformation fringes can be seen and a different number of fringes can be clearly observed. The deformation profile along the length of middle cantilever is plotted in Figures 2.22b and d respectively. This kind of analysis is useful when studying different parts of the same device simultaneously corresponding to the same input conditions.

2.4.2.2 MEMS Micro-Heater Analysis

Micro-heaters are basically resistive beams which can attain a temperature of 300–400 °C due to joule heating, when sufficient voltage is applied across them. The design of microheaters is optimized for low power consumption, low thermal mass, better temperature uniformity across the device and enhanced thermal isolation from the surroundings. The chip outlay of the MEMS micro-heater is shown in Figure 2.23a. Here an array of micro-heaters are fabricated, the single typical micro-heater is shown in Figure 2.23b. Here the thermal deformation in the electrodes (shown in the central dark color) appears because of the applied voltages between the electrodes.

The experimental set-up shown in Figure 2.9 is again used for a thermal deformation study of the micro-heater. The electrodes were connected to the power supply. A series of holograms were recorded corresponding to the different voltages between the electrodes. The optical microscopic image of the single micro-heater sample is shown in Figure 2.24a. The hologram of the same sample is recorded and the numerically reconstructed amplitude image of the hologram is shown in Figure 2.24b. Two points on different electrodes of the micro-heater as shown in Figure 2.24a are used to calculate the thermal deformation value at different voltages.

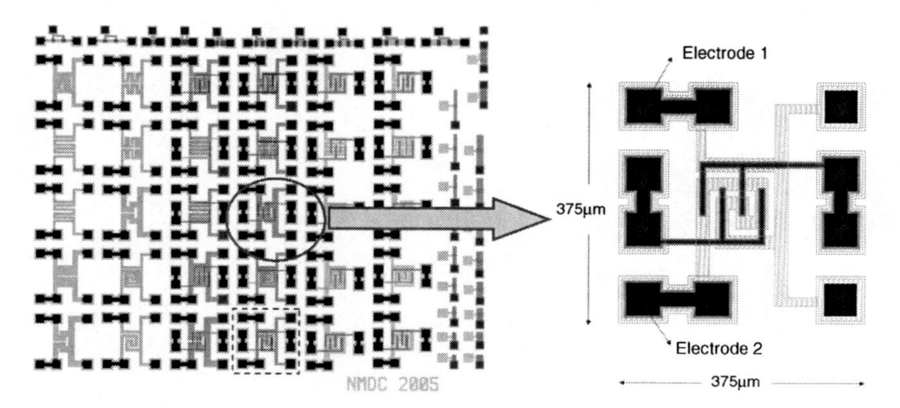

Figure 2.23 (a) Outlay of MEMS micro-heater chip; (b) Single micro-heater sample

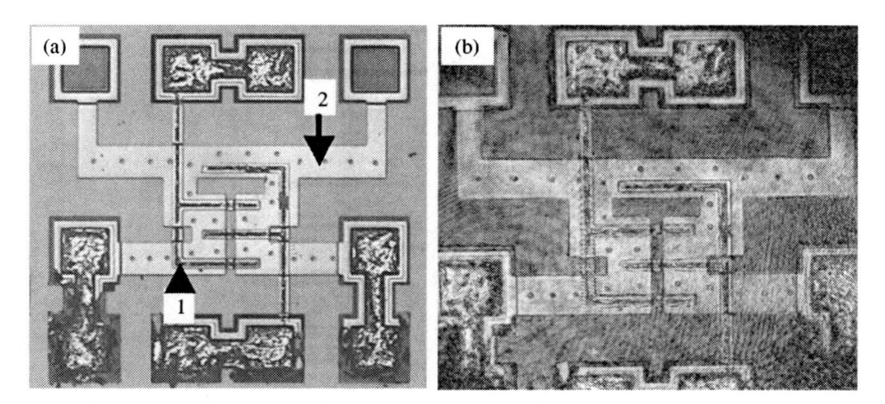

Figure 2.24 (a) Optical microscopic image of MEMS micro-heater at 5X magnification; (b) Digital holographic numerically reconstructed image

For dynamic thermal deformation measurements, the numerical reconstruction of the hologram is performed. The holograms were recorded by varying voltages from 1 volt to 10 volts. The phase data is reconstructed corresponding to each state. The subtraction of the phases of the deformed state (corresponding to the applied voltage) and the reference state (without applying any voltage) provides the deformation fringes. Figures 2.25a–l show the deformation fringes for the voltage changes from 1.0 volts to 3.7 volts. Deformation fringes can be clearly observed in the electrodes. As expected, the number of fringes increases with the increase in the voltage, which shows the increase in the thermal deformation with applied voltages.

The deformation fringes corresponding to the higher voltages are shown in Figures 2.26a–l, corresponding to the voltages varying from 7.0 volts to 9.7 volts. The number of fringes in the electrode is large, which shows the higher thermal deformation. It can be also observed that the deformation fringes are also expended in the other electrodes and increase with the increase in temperature. This shows the full field thermal deformation in the device at higher voltages. Thus, this kind of analysis is particularly useful for the full field study of smaller devices, where different components of the device show different deformation behavior, subject to the same input conditions.

The structural deformation due to thermal stress is quantitatively obtained from the phase analysis of the hologram fringes. The vertical deformation is measured at two points: one on the electrode, and the other on the microheater, as indicated in Figure 2.24a. The amplitude of deformation at the applied voltage can be calculated from the obtained fringe patterns shown in Figures 2.25 and 2.26. To do this, the phase difference value is converted into the path difference which represents the vertical deformation. Since an 8 bit CCD sensor is used to record the hologram and the wavelength of a source is

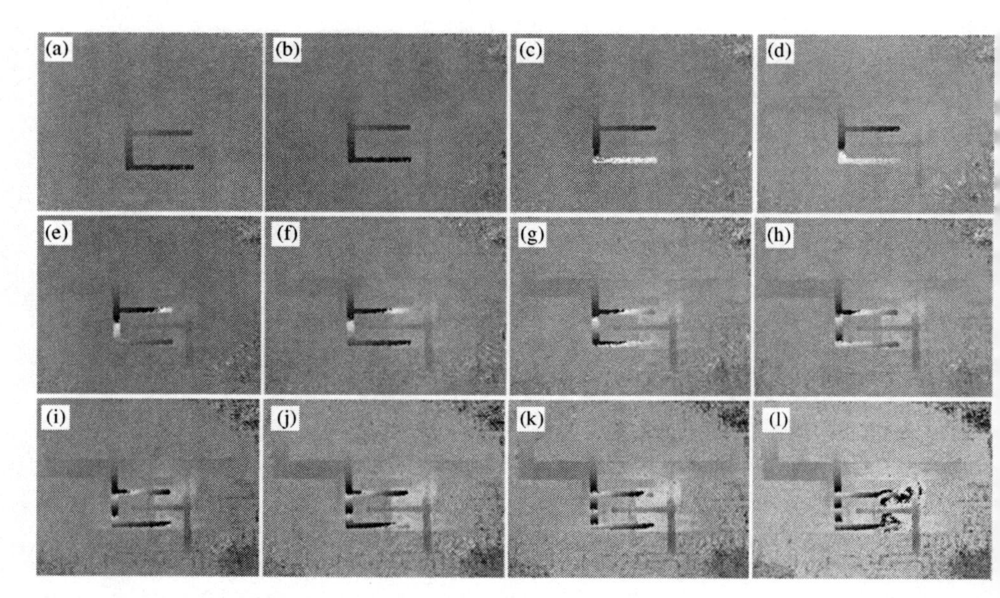

Figure 2.25 (a) Thermal deformation profile in MEMS micro-heater at applied voltages (a) 1.0 V; (b) 1.25 V; (c) 1.5 V; (d) 1.75 V; (e) 2.0 V; (f) 2.25 V; (g) 2.5 V; (h) 2.75 V; (i) 3.0 V; (j) 3.25 V; (k) 3.5 V; (l) 3.75 V (*Source*: (14) Reproduced by permission of © 2010 Institute of Physics (IOP) Publishing Ltd)

632.8 nm, it provides a theoretical vertical measurement accuracy of about 2.5 nm. The experimentally obtained deformations for the micro-heater and the electrode, as a function of applied power, are compared with the analytically calculated deflections of the doubly clamped beam (at the center of the beam) and the cantilever beam as shown in Figure 2.27. It is observed that the analytically obtained values closely match the experimentally observed deformations. As the electrode has a much lower spring constant (0.249 N m – 1) compared to the microheater structure (2.44 N m – 1), the vertical deflection of the electrode is much higher than the microheater. The proposed analysis is very useful in studying the response of different parts of the same microdevice simultaneously, under the same input conditions, and the method can be applied to characterize the MEMS structures. Thus, the thermo-mechanical characterization of MEMS-based microdevices is useful in inferring the residual stress inside the structure. Analysis of a deformation profile under operating conditions using a non-destructive approach is useful in evaluating the effect of a heating cycle on the structure and the effect of a fabrication process on the final behavior of the microheater in order to ensure the reliable fabrication of the microheater. The proposed method can also be extended to measure

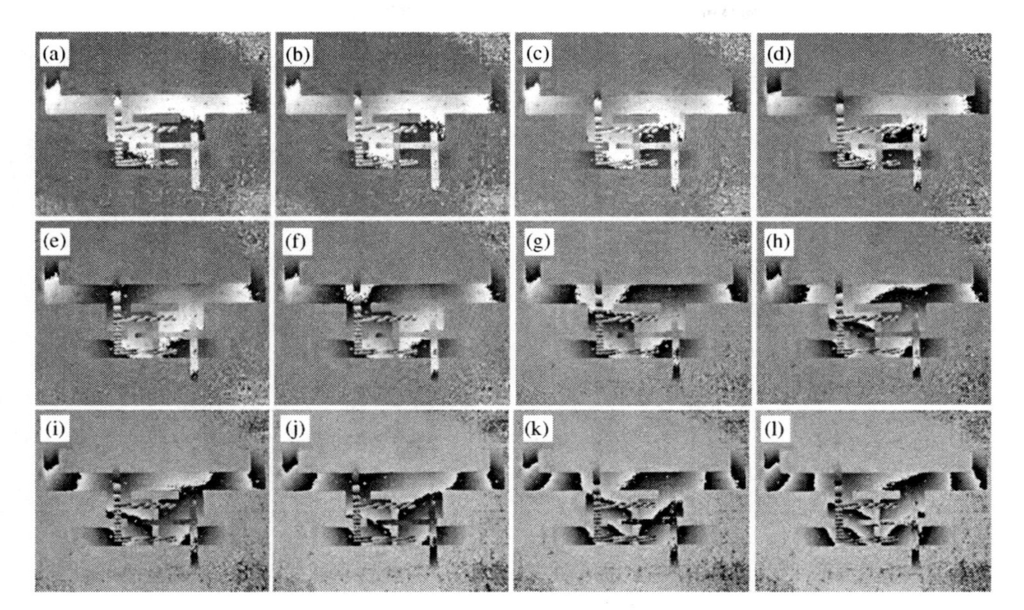

Figure 2.26 (a) Thermal deformation profile in MEMS micro-heater at applied voltages (a) 7.0 V; (b) 7.25 V; (c) 7.5 V; (d) 7.75 V; (e) 8.0 V; (f) 8.25 V; (g) 8.5 V; (h) 8.75 V; (i) 9.0 V; (j) 9.25 V; (k) 9.5 V; (l) 9.75 V (*Source*: (14) Reproduced by permission of © 2010 Institute of Physics (IOP) Publishing Ltd)

non-periodic deformation. In particular, applying this method of characterization, it is possible to evaluate the deformation profile of the MEMS structures not only under the static condition but also under the dynamic condition by taking a sequence of holograms using a high-speed CMOS camera.

2.4.3 Vibration Analysis

2.4.3.1 Study of Vibration Modes of a MEMS Diaphragm

Vibration analysis of an elliptical MEMS diaphragm with the major and minor axes 7 mm and 6.2 mm respectively is studied. The diaphragm was fabricated by bonding a piezoelectric plate onto a SOI (silicon on insulator) wafer with 20 μm thick device layer. The thickness of the piezoelectric layer was thinned down to about 40 μm by using chemical/mechanical polish. The back of the silicon was etched away by deep reactive ion etching. The diaphragm is excited by applying an AC driving voltage across the piezoelectric layer. The photograph of the MEMS diaphragm is shown in Figure 2.28.

An agilent 4294A precision impedance analyzer (40 Hz–110 MHz) was used to measure the impedance frequency spectrum of the diaphragm. The

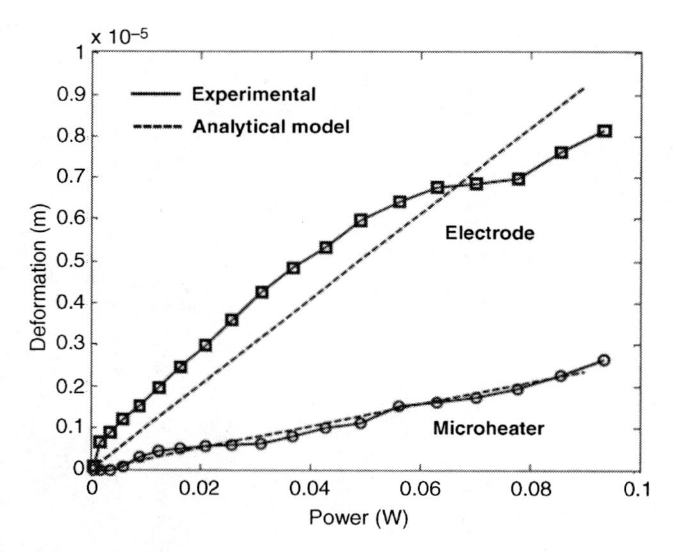

Figure 2.27 Experimentally measured and analytically computed thermal deformation with varying power at two different points on the device: (1) On the top electrode; (2) On the micro-heater (*Source*: (14) Reproduced by permission of © 2010 Institute of Physics (IOP) Publishing Ltd)

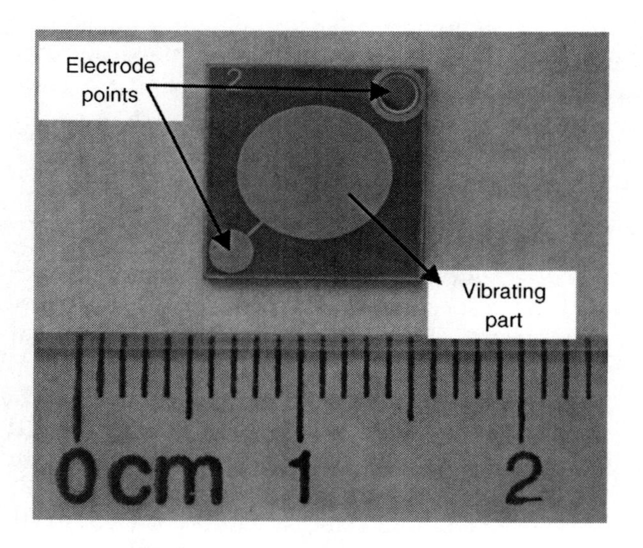

Figure 2.28 MEMS diaphragm

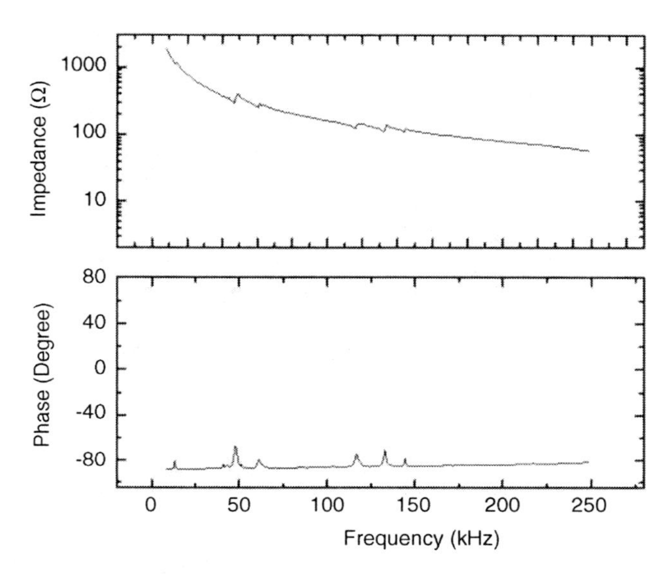

Figure 2.29 Impedance frequency spectrum of the MEMS diaphragm

frequency spectrum is shown in Figure 2.29. The response of the symmetric modes of the diaphragm shows the change in the impedance value, which represents a phase jump at the resonant frequencies. Thus, the phase jumps in Figure 2.29 show the resonant frequencies at the symmetric modes. However, the diaphragm also contains the anti-symmetric modes, but the impedance spectrum does not show the response corresponding to these resonant frequencies.

The digital reflection holography system is used to investigate the vibration analysis of the diaphragm. The mode shapes of the vibrating MEMS diaphragm are obtained from the reconstruction of time-averaged in-line holograms recorded corresponding to the resonant frequencies. The amplitude of the reconstructed real image wave, which is modulated by the J_0 function, gives the mode pattern. Figure 2.30 shows the vibration modes of the diaphragm corresponding to the resonant frequencies as shown in the frequency spectrum shown in Figure 2.29. These patterns can be obtained either by reconstruction of a single time-average hologram, or by subtraction of holograms in two states at the same frequency.

Time-averaged holograms were recorded corresponding to a wide range of applied frequencies. In addition to the frequencies shown in Figure 2.30, some additional resonant frequencies are also obtained. The mode shapes corresponding to these resonant frequencies are shown in Figures 2.31a–h. The vibration modes shown in Figure 2.31 can be identified as (a) (0,1) at 14 kHz; (b) (1,1) at 29 kHz; (c) (2,1) at 45 kHz; (d) (3,1) at 70 kHz; (e) (4,1) at 90 kHz,

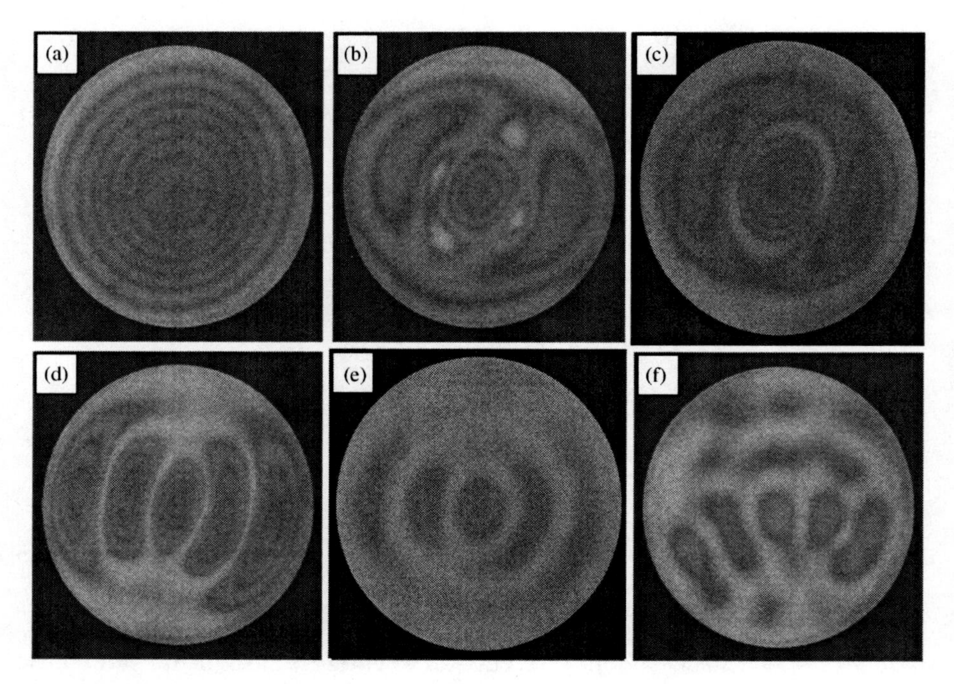

Figure 2.30 Vibration modes shapes of MEMS diaphragm verifying the frequency spectrum corresponding to the frequencies (a) 12 KHz; (b) 48 KHz; (c) 60 KHz; (d) 120 KHz; (e) 130 KHz; (f) 143 KHz (*Source*: (7))

(f) (1,2) at 105 kHz, (g) (6,1) at 145 kHz, and (h) (0,4) at 175 kHz, where the numbers in the parentheses refer to the torsional and bending modes. Some of antisymmetric modes can be clearly seen in Figures 2.31b, c and e.

The amplitude of vibration can be determined from the obtained mode patterns. Equation 2.32 is used for vibration amplitude measurements. The factor $\vec{K} \cdot \vec{z}_v = Kz_v g'$, where the geometrical factor g' needs to be calculated according to the system geometry, $K = 2\pi/\lambda$, for the experiment system $\lambda = 0.532\ \mu\text{m}$ and geometric factor $g' = 1.36$. This factor represents the zeros of the zero-order Bessel function of the reconstructed image. The zeros of first-order Bessel function used correspond to the dark fringes for calculating the amplitude values. The time-averaged fringes of the diaphragm from increasing the amplitude values at the resonant frequency 15 KHz are shown in Figure 2.32. The vibration amplitude values are plotted for different applied voltages and shown in Figure 2.33.

The vibration amplitude of the diaphragm is measured by the Laser Doppler vibrometer (LDV) system (by Polytech OFV-056C). The amplitude of

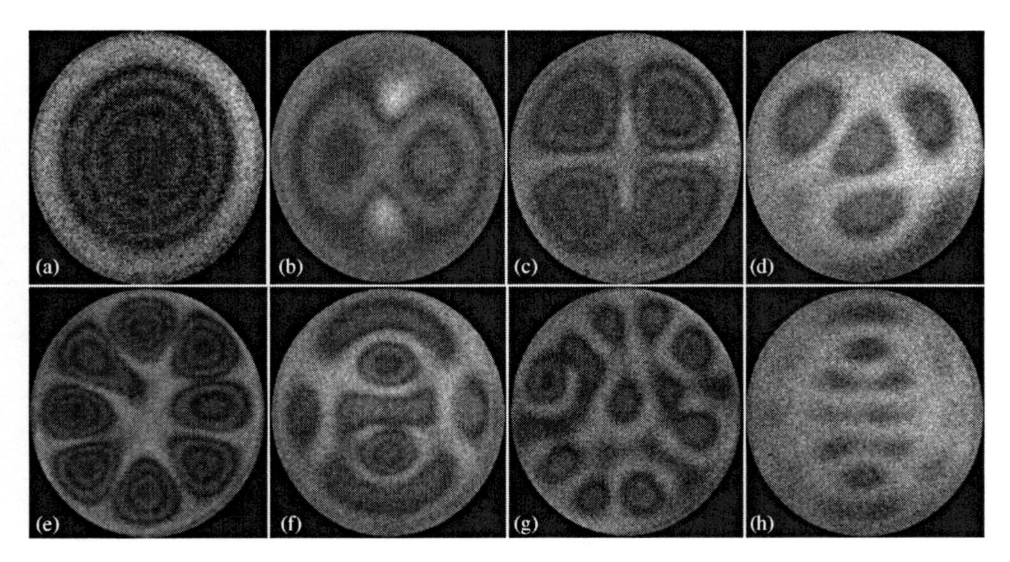

Figure 2.31 Vibration modes shapes corresponding to the resonant frequencies (a) 14 KHz; (b) 29 KHz; (c) 45 KHz; (d) 70 KHz; (e) 90 KHz; (f) 105 KHz; (g) 145 KHz; (h) 175 KHz (*Source*: [15] Reproduced by permission of © 2007 Elsevier)

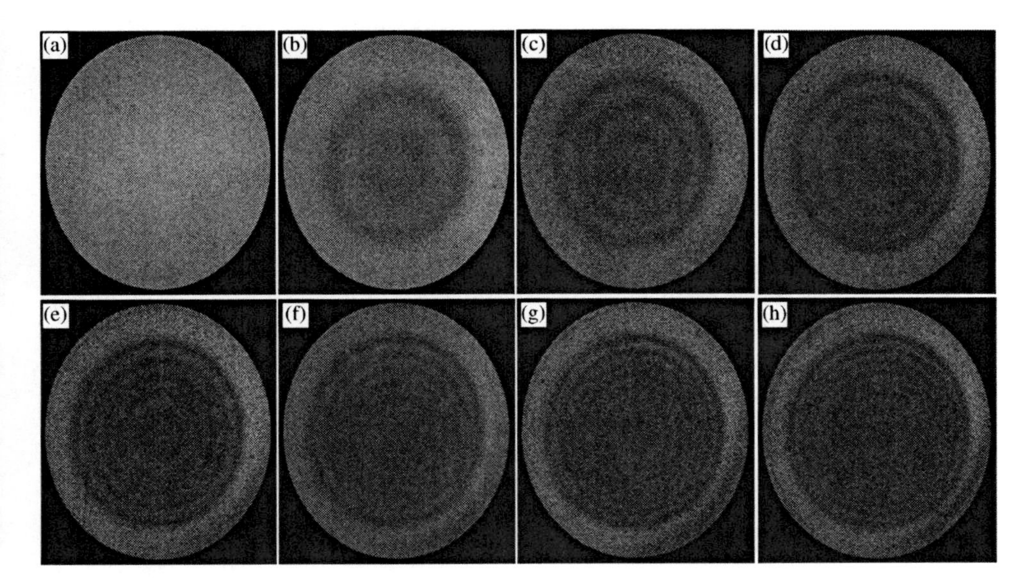

Figure 2.32 Vibration fringes of MEMS diaphragm as a function of applied voltages (a) 0 V; (b) 1.0 V; (c) 1.5 V; (d) 2.0 V; (e) 2.5 V; (f) 3.0 V; (g) 3.5 V; (h) 4.0 V

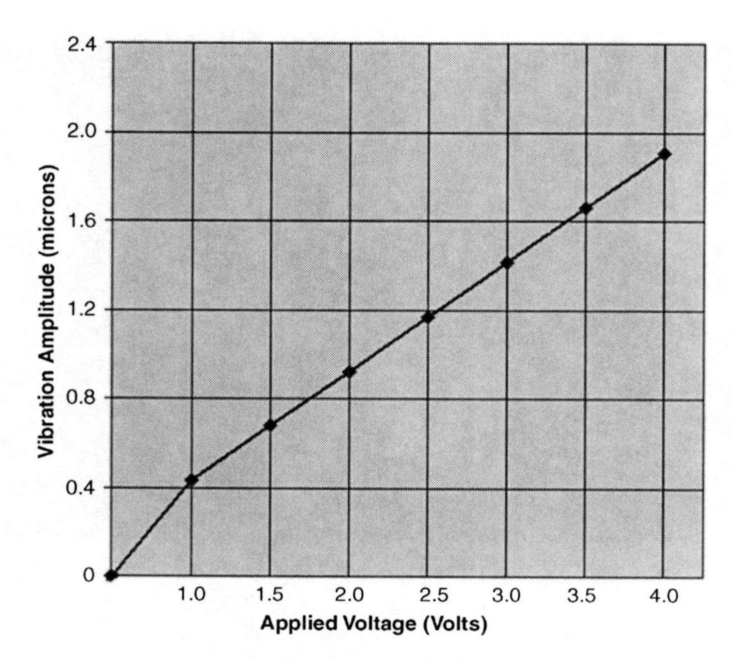

Figure 2.33 Vibration amplitude variation with applied voltage for a MEMS diaphragm

vibrations is measured at the center point of the diaphragm. The amplitude values on a few resonant frequencies are compared in Table 2.1.

Mean Deformation Analysis in the Presence of Vibrations

The main advantage in using digital holography is to obtain the quantitative phase data of the reconstructed image wave. As discussed previously, the numerical reconstructed phase from time-averaged holograms contains two parts, the first part represents the object surface roughness and the second part is called the time-average phase which shows the zeros of the J_0 function. For

Table 2.1 Comparison of vibration amplitude measurements

Frequency (KHz)	Vibration amplitude μm	
	LDV	DH
14	2.4386	2.275
32	0.5356	0.539
40	1.4400	1.321
44	1.2496	1.126

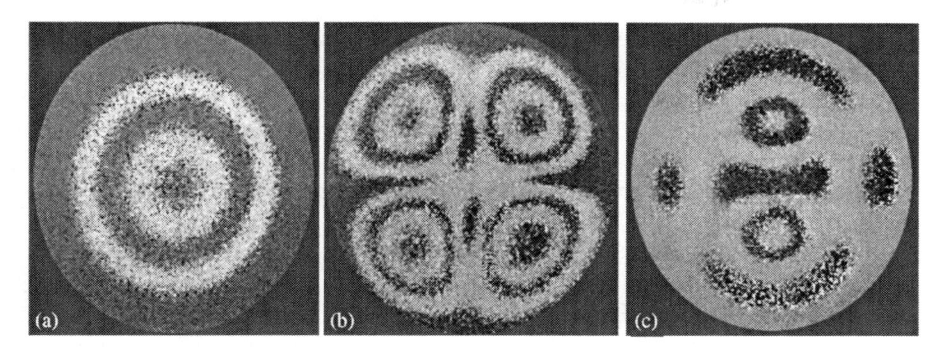

Figure 2.34 Time-average phase representing the binary jumps corresponding to the zero of the Bessel function (*Source*: (15) Reproduced by permission of © 2007 Elsevier)

the reconstruction of a single exposure hologram, the first part of the phase term reconstructs the object surface roughness data. Thus, for an object with optically rough surfaces, this contributes to the speckle noise. However, in the double exposure case, if there is any mean static deformation during the vibration amplitude change, the subtraction of the phase term represents the mean deformation. Again, for a rough object surface the mean deformation fringes occur due to speckle correlation, and for the time-averaged case, the mean static fringes mix with the vibration fringes. For the pure sinusoidal vibration of the object, the subtraction of phases of time-average and reference hologram provides only the time-average phase. This is shown in Figures 2.34a–c corresponding to the frequencies (a) 14 KHz; (b) 45 KHz; and (c) 105 KHz. Compared to Figure 2.35, all the zeros of the J_0 function can be clearly identified from the time-average phase. The eventual use of binary jumps is particularly

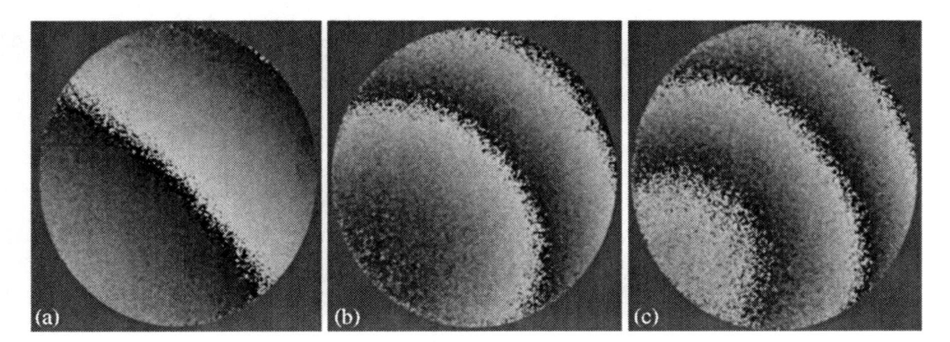

Figure 2.35 Modulo 2π interference phase representing mean static deformation obtained by applying offset voltage at non-resonant frequencies (*Source*: (15) Reproduced by permission of © 2007 Elsevier)

useful in the case of higher vibration amplitudes, because the time-averaged amplitude fringes are modulated by the zero-order Bessel function, so for higher-order fringes the contrast becomes significantly poor, while the time-average phase shows the binary jumps and thus has the same contrast for all orders.

Although the time-average phase contains binary values (with numerical values 0 and $\pm\pi$), it provides a clear representation of the zeros of J_0 function, while the simultaneous presence of the static phase appears as the speckle noise. The double exposure method cannot completely remove this noise because of the stochastic variations of speckles in the two exposures. This effect can be seen in Figure 2.35. In order to explore the mixing of the phase information during phase subtraction, first we have considered the case of pure static deformation which is achieved by selecting a non-resonant frequency and applying an offset voltage to the membrane between exposures. The phase subtraction represents modulo 2π interference phase, same as in conventional digital holographic interferometry. Figure 2.35 shows the pure mean static deformation fringes obtained from time-averaged in-line holograms. The holograms are recorded at a non-resonant frequency of 25 KHz, at a different voltage with increasing offset voltages (0.5 volt, 1.5 volts and 4.5 volts) applied by the frequency generator.

Mixing of the phase fringes is best visualized when the applied offset voltage excites the membrane in resonance. The mixing shows the cluster of mean deformation and time-average phases, which represents the exact vibration behavior of the membrane in the presence of mean static deformations. Thus the importance of mean static deformation is to study the actual behavior of vibrating objects in the presence of mean static change. Figures 2.36a–c show

Figure 2.36 Mixing of mean deformation and time average phases, representing the balancing of the diaphragm in the presence of both vibration and mean deformation (*Source*(15) Reproduced by permission of © 2007 Elsevier)

the patterns corresponding to the different mean and vibration amplitudes at a resonant frequency of 15 KHz. For double exposure recording, the reference hologram is recorded without vibration, and the time-averaged holograms recorded with driving voltages of 1.0, 1.5, and 2.0 volts and corresponding offset voltages 0.75, 1.5, and 2.5 volts respectively. It can be clearly seen in Figure 2.36a that the mean deformation fringe also appears inside the time-average phase with a phase jump. This phase jump may be attributed to the balancing condition of the membrane created by the offset voltage. As the offset voltage increases, the number of fringes inside the time average fringe also increases (Figures 2.36b and c).

2.4.3.2 Cantilever Vibration Analysis

Vibration analysis of the Aluminum nitride (AIN) cantilevers is performed using the time-averaged digital holographic microscopy system. AlN films have piezoelectric properties, and represent an alternative to PZT films. AIN cantilevers of size $800 \times 50\,\mu\text{m}$ long are fabricated using surface micromachining process [13]. The chip of the MEMS cantilevers device is attached to a standard SD card holder as shown in Figure 2.37, which is attached to the frequency generator. The lensless in-line digital holographic microscopy is

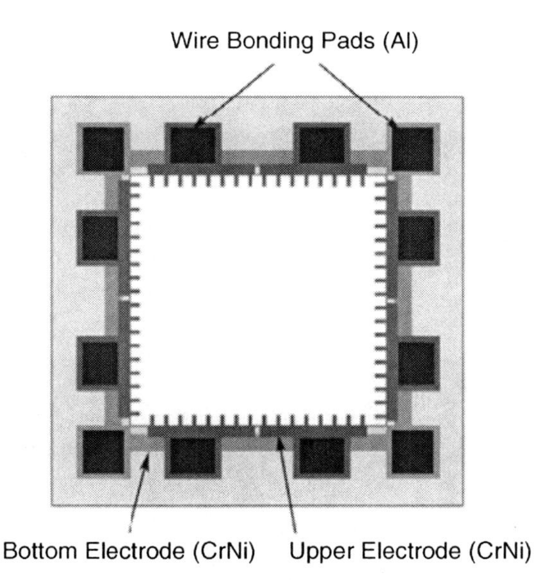

Figure 2.37 Layout of the $300 \times 50\,\mu\text{m}$ long cantilevers. The actuation of the cantilevers is done by applying a voltage between the upper and bottom metal electrodes (*Source*: (7))

Figure 2.38 Numerically reconstructed image of cantilevers (*Source*: (16) Reproduced by permission of © 2009 Optical Society of America)

explored for imaging and dynamic characterization of MEMS cantilevers. The numerically reconstructed image of the three static cantilevers electronically connected together is shown in Figure 2.38 and vibration analysis of these cantilevers is presented here.

The mode shapes of the vibrating cantilevers are obtained from the reconstruction of time-averaged in-line holograms. The amplitude of the reconstructed real image wave, which is modulated by the J_0 function, gives the mode pattern. Time-averaged holograms are recorded corresponding to the resonant frequencies of the cantilevers. The mode shapes corresponding to the first, second and third resonant frequencies are shown in Figure 2.39, which is obtained by amplitude reconstruction of time-averaged holograms. The vibration modes shown are presented corresponding to the first, second and third frequencies at (a) 30.46 kHz; (b) 191.40 kHz; and (c) 533.0 kHz respectively.

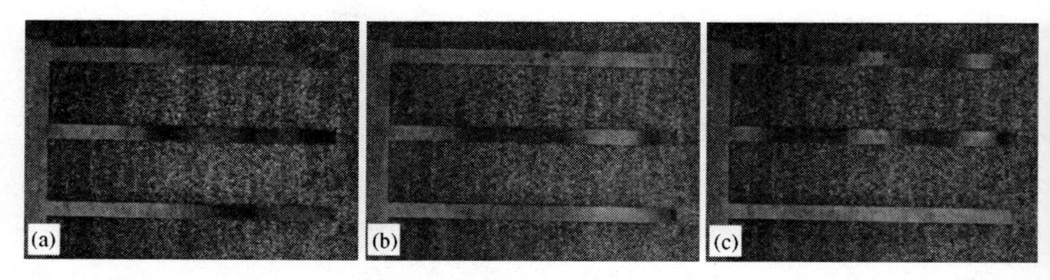

Figure 2.39 Mode shape corresponding to the resonant frequencies (a) 30.46 KHz; (b) 191.40 KHz; (c) 533.0 KHz (*Source*: (16) Reproduced by permission of © 2009 Optical Society of America)

The scanning of the frequencies corresponding to the resonant frequencies is performed and the amplitude of the cantilevers is calculated. For the experiment system $\lambda = 0.6328\,\mu\text{m}$ and geometric factor $g' = 2$. This factor represents the zeros of zero-order Bessel function of the reconstructed image. The zeros of first-order Bessel function used correspond to the dark fringes used to calculate the amplitude values. The frequency spectrum of the cantilevers is shown in Figure 2.40a. As can be clearly seen from Figure 2.39 that all the cantilevers are not vibrating at the same resonant frequency, this is further explored in Figures 2.40b–d. Figure 2.40b shows the resonant frequency of the each cantilever corresponding to the first resonant frequency and similarly Figures 2.40c and d show the second and third resonant frequencies respectively.

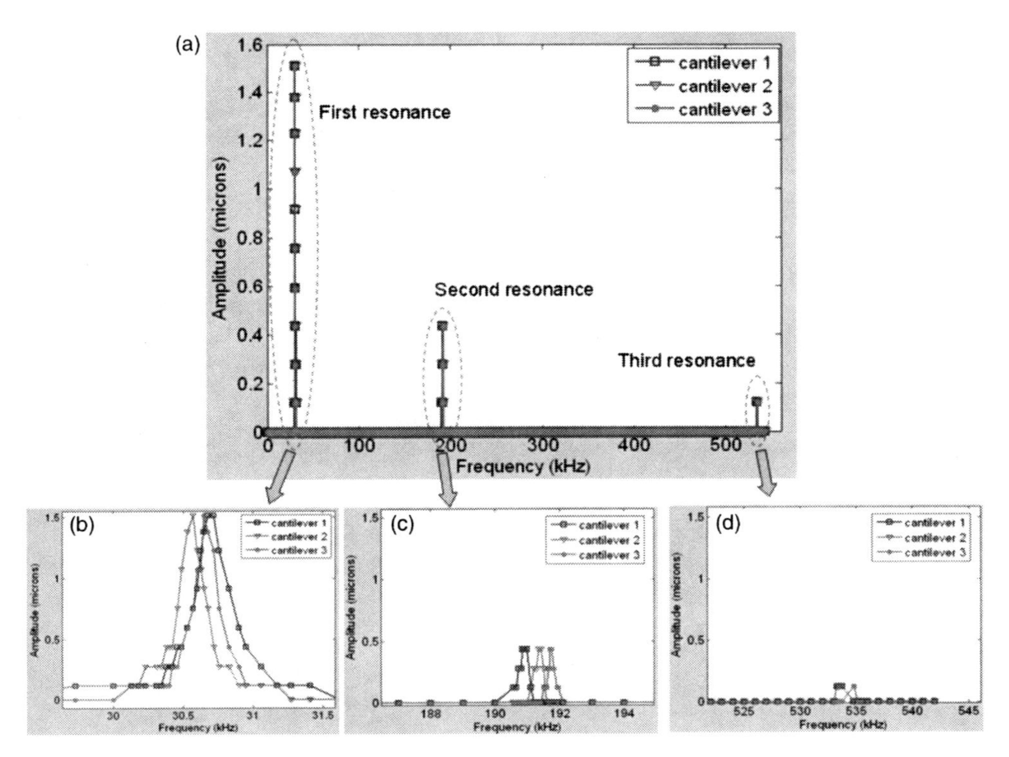

Figure 2.40 (a) Frequency spectrum of the cantilevers, and resonant frequencies of cantilevers corresponding to (b) First resonant; (c) Second resonant; (d) Third resonant frequency (*Source*: (16) Reproduced by permission of © 2009 Optical Society of America)

References

1. Gabor, D. (1949) Microscopy by reconstructed wavefronts. *Proc. Royal. Soc.*, **197**, 454–487.
2. Leith, E.N. and Upatnieks, J. (1964) Wavefront reconstruction with diffused illumination and three-dimensional objects. *J. Opt. Soc. Am.*, **54**, 1295.
3. Schnars, U. and Jüptner, W. (1994) Direct recording of holograms by a CCD target and numerical reconstruction. *Appl. Opt.*, **33**, 179.
4. Goodman, J.W. (1996) *Introduction to Fourier Optics*, McGraw-Hill, New York.
5. Schnars, U. and Jüptner, W.P.O. (2002) Digital holography and numerical reconstruction of holograms (REVIEW ARTICLE). *Meas. Sci. Technol.*, **13**, R85.
6. Xu, Lei, Peng, Xiaoyuan, Miao, Jianmin, and Asundi, Anand (2001) Studies of digital microscopic holography with applications to microstructure testing. *Appl. Opt.*, **40**, 5046.
7. Singh, V.R. (2008) Lensless inline digital holography for dynamic applications. Ph. D. thesis, Nanyang Technological University, Singapore.
8. Asundi, A. and Singh, V.R. (2009) Digital holography for MEMS application. *Novel Techniques in Microscopy*, OSA Technical Digest (CD) (Optical Society of America, 2009), paper JTuA2.
9. Asundi, A. and Singh, V.R.(Publication date: 09 April 2009) ''Holographic microscope and a method of holographic microscopy'', *US Patent Pending*, Application number: 20090091811.
10. Singh, V.R., Lui, S., and Asundi, A. (2010) Digital reflection holography based systems for MEMS measurements. *Proc. SPIE*, **7718**, 77180.
11. Singh, V.R., Lui, S., and Asundi, A. (2009) Compact handheld digital holographic microscopy system development. *Proc. SPIE*, **7522**, 75224.
12. Chee, O.C., Singh, V.R., Sim, E., and Asundi, A. (2008) Development of simple user-friendly commercial digital holographic microscope. *Proc. of SPIE*, **6912**, 69120V–1.
13. Asundi, A. and Singh, V.R. (2006) Time averaged in-line digital holographic interferometry for vibration analysis. *Appl. Opt.*, **45**, 2391.
14. Jayaraman, B., Singh, V.R., Asundi, A. *et al.* (2010) Thermo-mechanical characterization of surface-micromachined microheaters using in-line digital holography. *Meas. Sci. Technol.*, **21**, 015301 (10pp).
15. Singh, V.R., Asundi, A., Miao, J. *et al.* (2007) Dynamic characterization of MEMS diaphragm using time averaged in-line digital holography. *Opt. Comm.*, **280**, 285.
16. Singh, V.R. and Asundi, A. (2009) In-line digital holography for dynamic metrology of MEMS. *Chinese Optics Letters*, **7**, 1117.

3

Digital Transmission Holography and Applications

Qu Weijuan
Center of Innovation, Ngee Ann Polytechnic, Singapore

3.1 Historical Introduction

In 1948, Dennis Gabor introduced the concept of holography [1] for wavefront reconstruction in order to improve the resolution of electronic microscopes. In his original description, holography is a two-step method of optical imaging. In the first step, the object is illuminated with a coherent monochromatic wave, and the diffraction pattern resulting from the interference of the coherent secondary wave issuing from the object with the strong, coherent background is recorded on a photographic plate. In the second step, the photographic plate, suitably processed, is replaced in the original position and illuminated with the coherent background alone. The coherent secondary wave is reconstructed together with an equally strong "twin wave" which has the same amplitude, but opposite phase shifts relative to the background. Though the distortion from the twin image strictly limited the application of this holographic method in the early days, it still found many applications in measuring small objects. The utility of in-line holography as an experimental tool for particle studies was demonstrated by Thompson [2]. Then it was widely

Digital Holography for MEMS and Microsystem Metrology, First Edition. Edited by Anand Asundi.

applied in particle sizing [3] and particle coordinate [4] and velocity [5] measurements. At first, the holograms were called Fraunhofer or far field holograms [6], because the hologram was recorded at a distance from the object that was effectively in the far field of the small particles. When a spherical diverging wavefront is used in hologram recording, the recorded particles have a magnification that depends on their location. This proved to be a limitation on the depth of field of the optical reconstruction process, thus people prefer to use the plane wave for recording and reconstruction, because the magnification of the reconstructed particles will be uniform over the entire recorded volume [7].

At the same time, people found many optical methods to suppress the twin image and the background influence. Gabor and Goss [8] designed an interference microscope in which two quadrature-phase holograms were recorded and reconstructed to suppress the conjugate image. However, the construction of this interference microscope was too complicated for practical use. The scheme of off-axis holography [9] was designed by employing a beam splitter [10] or introducing additional lenses into the recording and reconstructing process [11] to remove the twin image.

The undulatory nature of light and the concept of coherence are important for a better understanding of image formation in optics. Holography can provide 3D data in only one shot. The discovery of long coherence sources made a wide range of holographic interferometry a practical tool in metrology. The development of computer technology allowed the transfer of the reconstruction process to the computer. Numerical reconstruction of the optically recorded hologram was initiated by Goodman and Lawrence and developed by Yaroslavski *et al.* [12]. They sampled optically enlarged parts of in-line and Fourier holograms recorded on a photographic plate. These digitized conventional holograms were reconstructed numerically. Onural and Scott [13] improved the reconstruction algorithm and applied it to particle measurement. Then the numerical reconstruction algorithm was widely developed to suppress the twin image [14]. Direct recording of Fresnel holograms with a charged coupled devices (CCD) [15] enables the full digital recording and the numerical processing of holograms. Digital holography comes into being and plays an important role in quantitative phase measurement.

Digital holography enables the direct extraction of phase data from the recorded light waves. The well-known phase shifting method and numerical reconstruction algorithm for in-line holography are presented by Yamaguchi [16]. The phase of the twin image is successfully removed to achieve the phase of the real image. This process was soon applied to three-dimensional microscopy [17], encryption [18], pattern recognition and surface contouring [19]. The drawback of this method is the need to record multiple holograms which is not suitable in real-time dynamic phase monitoring. Since the mid-1990s, off-axis digital holographic interferometry (double exposure) has been developed, improved and applied to many measurement tasks such as deformation analysis, shape measurement, particle tracking, and refractive index distributions measurement [20] in transparent media, due to temperature or concentration variations and bio-imaging application. The limited sampling capacity of the electronic camera is the reason why different approaches to achieve microscopic imaging with digital holography have been sought, and thus revives digital holographic microscopy (DHM) [21]. DHM, with the introduction of a microscope objective (MO), gives very satisfactory measurement results both in lateral resolution and vertical resolution. Nevertheless, the MO introduces a phase curvature to the object wave which increases the complexity of the numerical reconstruction process. For a long time, double exposure was used to remove the additional phase curvature to give a relative phase measurement of the test object because of the lack of a powerful digital phase compensation procedure. It is the most reliable approach that is immune from aberrations, but again it needs two holograms.

In 1999, Cuche *et al.* proposed that not only amplitude but also phase could be extracted from a single digital hologram [22]. The introduction of the concept of the digital reference wave that compensates the role of the reference wave in off-axis geometry has successfully removed the off-axis tilt of the optical system. Since then, the numerical reconstruction method for the digital hologram has been further developed [23–25].

The most two successful reconstruction methods are the Fresnel transform method [26] and the angular spectrum method. Different mathematical descriptions of wave propagation are used in the two methods. Generally, numerical phase compensation is needed for both methods as quantitative phase measurement is involved. In off-axis digital holography, numerical phase compensation is needed because of the off-axis tilt. In DHM, it is needed because the use of a microscope objective (MO) affects the divergence of the object wave and results in a wavefront aberration between the object and reference waves.

Examples, such as the Linnik interferometer [27] and the Mach-Zehnder interferometer [28, 29], can physically solve the problem by introducing the same curvature in the reference wave. However, this requires precise alignment of all the optical elements involved. Previously, it was difficult to achieve the correct compensation. Now it is easy with the help of real-time monitoring software. In this chapter we will introduce the details of physically spherical phase compensation of transmission digital holographic microscopy.

3.2 The Foundation of Digital Holography [30]

The difference between classical holography and digital holography is in the recording material and reconstruction processing of the hologram. For classical holography, a hologram is recorded by photographic plate and then chemically processed. The original recording reference wave is generally required in the optical reconstruction. For digital holography, a hologram is recorded with an electronic device and stored in the computer memory. Current or later numerical reconstruction can be performed by using the digital replica of the optical reference wave. However, as a wavefront recording and reconstruction method, they have the same foundation which is the diffraction and interference of optical wave fields. Understanding of a "wavefront" is necessary in digital holography for a more thorough understanding of the physics involved.

As a laser is used as the source of digital holography, we shall restrict our attention almost entirely to monochromatic, linearly polarized wave fields, that is, wave fields that can be adequately described by a real-value scalar function of position. We define the complex amplitude of this wave field as:

$$u(x,y,z) = a(x,y,z)\exp\left[j\varphi(x,y,z)\right]. \tag{3.1}$$

When a wave field impinges on a surface normally, the rate per unit area at which radiant energy arrives at the surface is referred to as the irradiance of the wave field at that surface. Detectors respond to such quantities which is proportional to the square modulus of the complex amplitude for a monochromatic wave field. The irradiance is denoted by $I(x, y, z)$ and given as:

$$I(x,y,z) = |u(x,y,z)|^2. \tag{3.2}$$

As is well known, holography not only can record the intensity but also the phase of the wavefront, but it cannot directly provide the data. One needs to further process the hologram to obtain the intensity and phase data. This further process does not mean a chemical process, but a numerical process of the digital hologram.

The aim of digital holography is the employment of CCD arrays to record holograms which are then stored in a computer memory and can be reconstructed numerically. Since the CCD arrays have $N \times N$ light-sensitive pixels, the digitized wavefront which is captured can be denoted by an $N \times N$ matrix, as shown in Figure 3.1. Thus the operation of the digital hologram is actually the operation between different matrixes in mathematics. The mathematical description of optical wave fields is then important for digital holography.

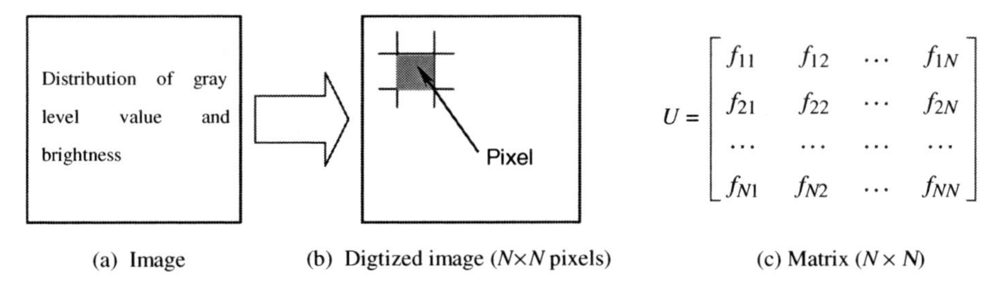

Figure 3.1 Digitization and matrix denotation of the image. (a) Image; (b) Digitized image ($N \times N$ pixels); (c) Matrix ($N \times N$)

Plane Wave Fields

The complex amplitude description of a general monochromatic plane wave field propagating along the k direction in a rectangular coordinated system is

$$u(x, y, z) = A \exp\left[jk(x \cos \alpha + y \cos \beta + z \cos \gamma) + j\Phi\right] \quad (3.3)$$

where A is the strength, or amplitude, of the wave field, Φ is its phase at the origin, and ($\cos\alpha$, $\cos\beta$, $\cos\gamma$) are the direction cosines of its propagation vector. As usual, $k = 2\pi/\lambda$, where λ denotes the wavelength of the light. From the relationship,

$$\cos^2\alpha + \cos^2\beta + \cos^2\gamma = 1 \quad (3.4)$$

we can write:

$$\begin{aligned} u(x, y, z) = & A \exp(jkz \cos \gamma + j\Phi) \exp\left[jk(x \cos \alpha + y \cos \beta)\right] \\ & A \exp\left(jkz\sqrt{1 - \cos^2\alpha - \cos^2\beta} + j\Phi\right) \exp\left[jk(x \cos \alpha + y \cos \beta)\right] \end{aligned} \quad (3.5)$$

In any plane for which z is constant, say $z = z_1$, Equation 3.5 may be written as:

$$u(x, y, z_1) = A \exp\left(jkz_1\sqrt{1 - \cos^2\alpha - \cos^2\beta}\right) \exp(j\Phi) \exp\left[jk(x \cos \alpha + y \cos \beta)\right] \quad (3.6)$$

This expression describes the x and y dependence of the complex amplitude in the plane $z = z_1$, and the quantity $\exp\left(jkz_1\sqrt{1 - \cos^2\alpha - \cos^2\beta}\right)$ is simply a complex constant that accounts for the phase difference between any point in the $z = 0$ plane and the corresponding point in the plane $z = z_1$. In other words, the difference in phase between the points (x, y, 0) and (x, y, z_1) is $\exp\left(jkz_1\sqrt{1 - \cos^2\alpha - \cos^2\beta}\right)$ for all x and y.

As long as we are interested in describing a plane wave field only in the $z = z_1$ plane, we can set the constant $\exp\left[j\left(kz_1\sqrt{1-\cos^2\alpha-\cos^2\beta}+\Phi\right)\right]$ equal to unity. Then Equation 3.6 is simplified as:

$$u(x, y, z_1) = A\exp\left[jk(x\cos\alpha + y\cos\beta)\right] \tag{3.7}$$

Generally, $\exp[jk(x\cos\alpha + y\cos\beta)]$ is called the linear phase factor of the plane wave fields. If this factor is included in the expression of the complex amplitude distribution of the plane, it shows that there is one plane wave in direction $\cos\alpha$, $\cos\beta$ going through the plane.

The digitized wavefront of plane wave field is:

$$u\,(m\Delta x, n\Delta y) = A\exp\left[j\frac{\lambda}{2\pi}(m\Delta x\cos\alpha + n\Delta y\cos\beta)\right] \tag{3.8}$$

where m and n are $M \times N$ matrixes, they denote the address of the sampling pixel in the x and y direction respectively, Δx and Δy are the sampling pixel size in x and y direction respectively. Consequently u is the $M \times N$ matrix. The phase can be calculated as:

$$\varphi(m\Delta x, n\Delta y) = \arctan\left(\frac{\mathrm{Im}[u(m\Delta x, n\Delta y)]}{\mathrm{Re}[u(m\Delta x, n\Delta y)]}\right) \tag{3.9}$$

It can be numerically displayed by the computer as shown in Figure 3.2.

In free space, the nature of a plane wave field does not change as it propagates; a plane wave field at $z = z_1$ is also a plane wave field at any other

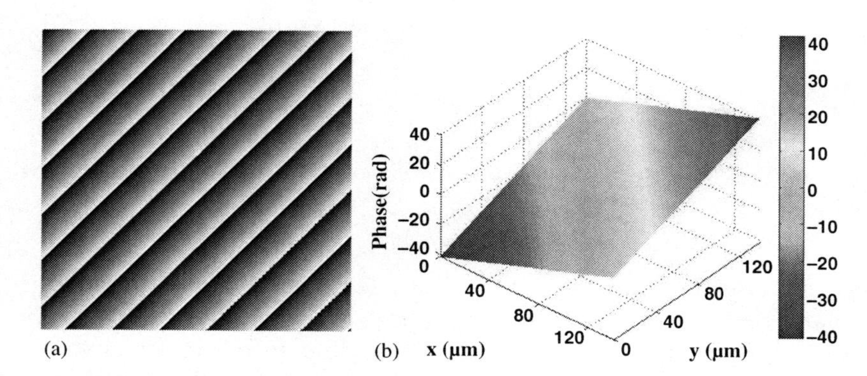

Figure 3.2 Phase of the wavefront for a plane wave field propagating at an angle 45° to the x and y direction. (a) Phase modulo by 2π; (b) 3D quantitative phase

plane $z = z_i$, and the only difference in the wave fields at these two planes is in the phase. With regard to digital holography, the off-axis geometry in the recording process has a tilt effect on the reconstructed wavefront. This off-axis tilt can be removed by the application of a numerical reference plane wave field in the numerical reconstruction.

Spherical Wave Field

We next investigate the behavior of spherical wave fields. To do so, we place a point source of light at the point $(S_x, S_y, 0)$ and specify the complex amplitude of the resulting wave field in the region $z > 0$. Light emitted from a point source has a diverging spherical wavefront. Generally, in the $z = z_1$ plane, the complex amplitude description of monochromatic diverging spherical wave field in quadratic-phase approximations is:

$$u(x, y) = A \exp\left\{-j\frac{\pi}{\lambda z_1}\left[(x - S_x)^2 + \left(y - S_y\right)^2\right]\right\} \tag{3.10}$$

where A is the unit amplitude of the wave field, and λ is the wavelength.

The digitized wavefront of diverging spherical wave field is:

$$u(m\Delta x, n\Delta y) = A \exp\left[-j\frac{\pi}{\lambda z_1}\left((m\Delta x - S_x)^2 + \left(n\Delta y - S_y\right)^2\right)\right] \tag{3.11}$$

where m and n are $M \times N$ matrixes, they are the same as those in Equation 3.8 where they denote the address of the sampling pixel in the x and y directions respectively, Δx and Δy are the sampling pixel sizes in the x and y directions respectively. As was the case of the plane wave field, the nature of a spherical wave field in free space does not change as it propagates; its amplitude and phase may vary with position, as well the curvature of its wavefronts, but it remains a spherical wave. The phase can be calculated by Equation 3.9 and numerically displayed by the computer as shown in Figure 3.3.

The digitized wavefront of a converging spherical wave field to the point (S_x, S_y, z_1) is:

$$u(m\Delta x, n\Delta y) = A \exp\left[j\frac{\pi}{\lambda z_1}\left((m\Delta x - S_x)^2 + \left(n\Delta y - S_y\right)^2\right)\right] \tag{3.12}$$

The phase calculated by Equation 3.12 is numerically displayed by the computer as shown in Figure 3.4.

In digital holographic microscopy, owing to the use of the microscope objective (MO), a diverging spherical wavefront must be taken into account for the phase measurement.

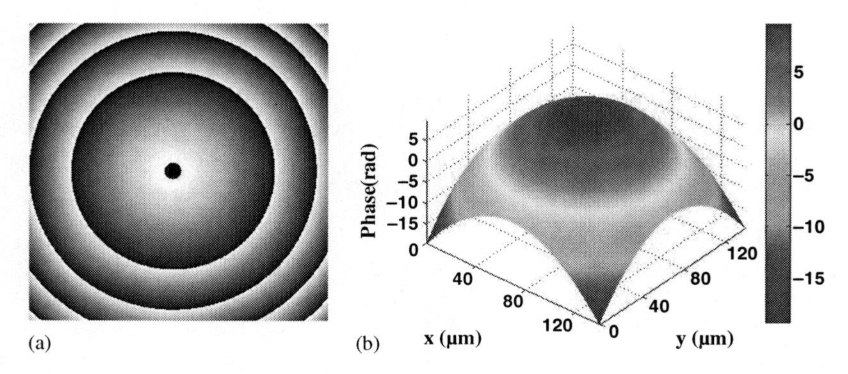

Figure 3.3 Phase of the wavefront for a diverging wave field (a) Phase modulo by 2π; (b) 3D quantitative phase

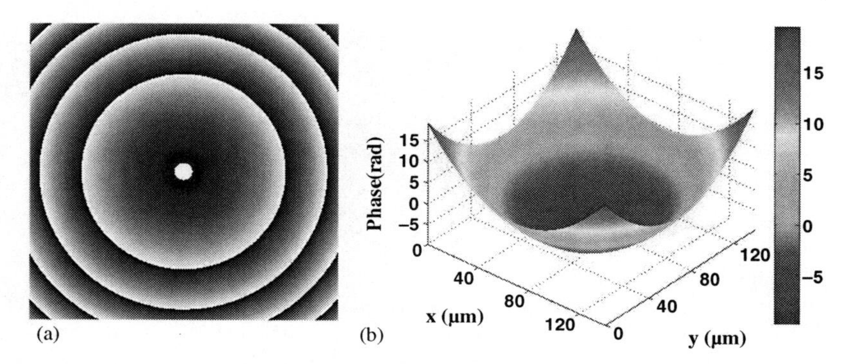

Figure 3.4 Phase of the wavefront for a converging wave field (a) Phase modulo by 2π; (b) 3D quantitative phase

3.2.1 Theoretical Analysis of Wavefront Interference

Each optical wave field consists of an amplitude distribution as well as a phase distribution, but all detectors only register intensities. The recorded intensity is modulated by phases of the involved interference wavefronts. Actually interference is the way of phase recording. It goes back to the beginning of holography. A hologram is created by the interference, that is, an illuminating wavefront modulated by an unknown wavefront coming from the object, called the object wave O, is added to the reference wave R to give an intensity modulated by their phases. The intensity $I_H(x, y)$ of the sum of two complex fields can be written as:

$$I_H(x, y) = |O + R|^2 = |O|^2 + |R|^2 + RO^* + R^*O \tag{3.13}$$

where RO^* and R^*O are the interference terms with R^* and O^* denoting the complex conjugate of the two waves.

Thus, if

$$O(x,y) = A_O \exp[-j\,\varphi_O(x,y)] \tag{3.14a}$$

$$R(x,y) = A_R \exp[-j\,\varphi_R(x,y)] \tag{3.14b}$$

where, A_O and A_R are the constant amplitude of the object wave and reference wave, respectively; $\varphi_O(x, y)$ and $\varphi_R(x, y)$ are the phase of the object wave and reference wave, respectively. The intensity of the sum is given by:

$$I_H(x,y) = A_O^2 + A_R^2 + A_O A_R \cos[\varphi_R(x,y) - \varphi_O(x,y)]. \tag{3.14c}$$

It is obvious that the intensity depends on the relative phases of the two involved interference waves. In order to get sufficient information for the reconstruction of the phase of the test object, one needs to specify the detailed characters of the involved waves.

3.2.1.1 Interference between Two Plane Waves

Two plane waves are digitized as:

$$O(m\Delta x, n\Delta y) = A_O \exp\left[j\frac{\lambda}{2\pi}(m\Delta x \cos\alpha_O + n\Delta y \cos\beta_O)\right] \tag{3.15a}$$

$$R(m\Delta x, n\Delta y) = A_R \exp\left[j\frac{\lambda}{2\pi}(m\Delta x \cos\alpha_R + n\Delta y \cos\beta_R)\right] \tag{3.15b}$$

where $(\cos\alpha_O, \cos\beta_O)$ is the direction of the object wave; $(\cos\alpha_R, \cos\beta_R)$ is the direction of the reference wave. The intensity is:

$$\begin{aligned} I_H(x,y) &= |O+R|^2 \\ &= A_O^2 + A_R^2 + A_O A_R \cos\left\{\frac{\lambda}{2\pi}[m\Delta x(\cos\alpha_R - \cos\alpha_O) + n\Delta y(\cos\beta_R - \cos\beta_O)]\right\} \end{aligned} \tag{3.15c}$$

Because there is no other test object than the object wave itself, the pattern of the hologram obtained from the interference between two plane waves is always a straight fringe. The direction of the fringes is decided by $[(\cos\alpha_O\text{-}\cos\alpha_R), (\cos\beta_O\text{-}\cos\beta_R)]$. As shown in Figure 3.5, if the angle between the two waves

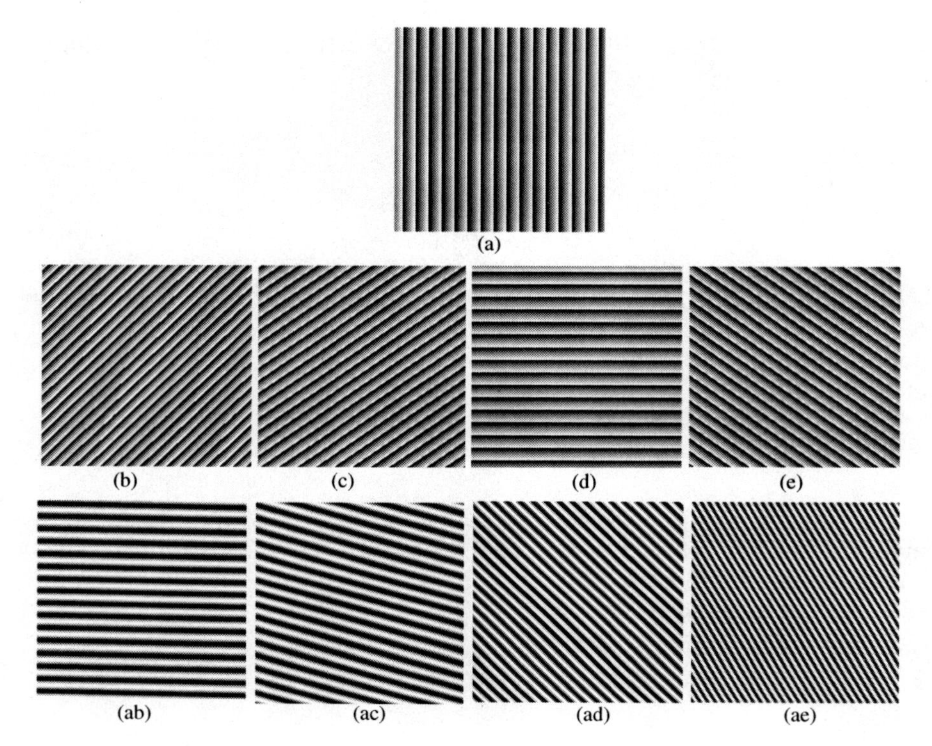

Figure 3.5 Interference pattern of two plane wavefronts. Phase of the plane wavefront modulo by 2π (a), (b), (c), (d) (e); The interference fringe pattern of the corresponding plane wavefronts (ab), (ac), (ad), (ae)

changes, there will be a corresponding direction and period change for the straight fringes.

3.2.1.2 Interference between a Plane Wave and a Spherical Wave

A diverging spherical wavefront from the point source (S_{Ox}, S_{Oy}, z_d) is digitized as:

$$O_d(m\Delta x, n\Delta y) = A_{Od} \exp\left[-j\frac{\pi}{\lambda z_d}\left((m\Delta x - S_{Ox})^2 + \left(n\Delta y - S_{Oy}\right)^2\right)\right] \quad (3.16a)$$

where A_{Od} is the unit amplitude of the diverging spherical wave; z_d is the distance between the point source and the observation plane. Interference with the plane reference wavefront is expressed by Equation 3.16b. The hologram is observed in a plane which is normal to the propagation direction of the diverging spherical wavefront. The intensity is:

$$I_H(x,y) = |O_d + R|^2$$
$$A_{Od}^2 + A_R^2 + A_{Od}A_R\cos\left[\frac{\pi}{\lambda z_d}\left((m\Delta x - S_{Ox})^2 + (n\Delta y - S_{Oy})^2\right)\right.$$
$$\left. + \frac{\lambda}{2\pi}(m\Delta x \cos\alpha_R + n\Delta y \cos\beta_R)\right] \quad (3.16b)$$

A converging spherical wavefront to the point (S_{Ox}, S_{Oy}, z_c) is digitized as:

$$O_c(m\Delta x, n\Delta y) = A_{Oc}\exp\left[j\frac{\pi}{\lambda z_c}\left((m\Delta x - S_{Ox})^2 + (n\Delta y - S_{Oy})^2\right)\right] \quad (3.17a)$$

where A_{Oc} is the unit amplitude of the converging spherical wave; z_c is the distance between the point source and the observation plane. Interference with the plane reference wavefront is expressed by Equation 3.17b. The hologram is observed in a plane which is normal to the propagation direction of the converging spherical wavefront. The intensity is:

$$I_H(x,y) = |O_c + R|^2$$
$$= A_{Oc}^2 + A_R^2 + A_{Oc}A_R\cos\left[\frac{\pi}{\lambda z_c}\left((m\Delta x - S_{Ox})^2 + (n\Delta y - S_{Oy})^2\right)\right.$$
$$\left. - \frac{\lambda}{2\pi}(m\Delta x \cos\alpha_R + n\Delta y \cos\beta_R)\right] \quad (3.17b)$$

It is noticed that there are both square and linear terms of Δx and Δy in the cosine function of Equations 3.16 b and 3.17b. Consequently we separately draw the fringe pattern of the cosine function of the square term of Δx and Δy and the linear term of Δx and Δy in order to see the difference between them, as shown in Figure 3.6. Figure 3.6a shows the fringe pattern of $\cos[\frac{\lambda}{2\pi}(m\Delta x \cos\alpha_R + n\Delta y \cos\beta_R)]$, the linear factor. It is straight fringes in direction ($\cos\alpha_R$, $\cos\beta_R$). Figure 3.6b shows the fringe pattern of $\cos[\frac{\pi}{\lambda z_c}((m\Delta x - S_{Ox})^2 + (n\Delta y - S_{Oy})^2)]$, the spherical factor. This is circular fringes. Figure 3.6c shows the fringe pattern of Equation 3.16b. Figure 3.6d shows the fringe pattern of Equation 3.17b. It seems that Figures 3.6c and d are different combinations of Figures 3.6a and b. Since the fringe patterns are coming from diverging spherical wavefronts and converging spherical wavefronts interfere with the same plane wavefront, the difference in hologram patterns is an indication of the spherical wavefront property.

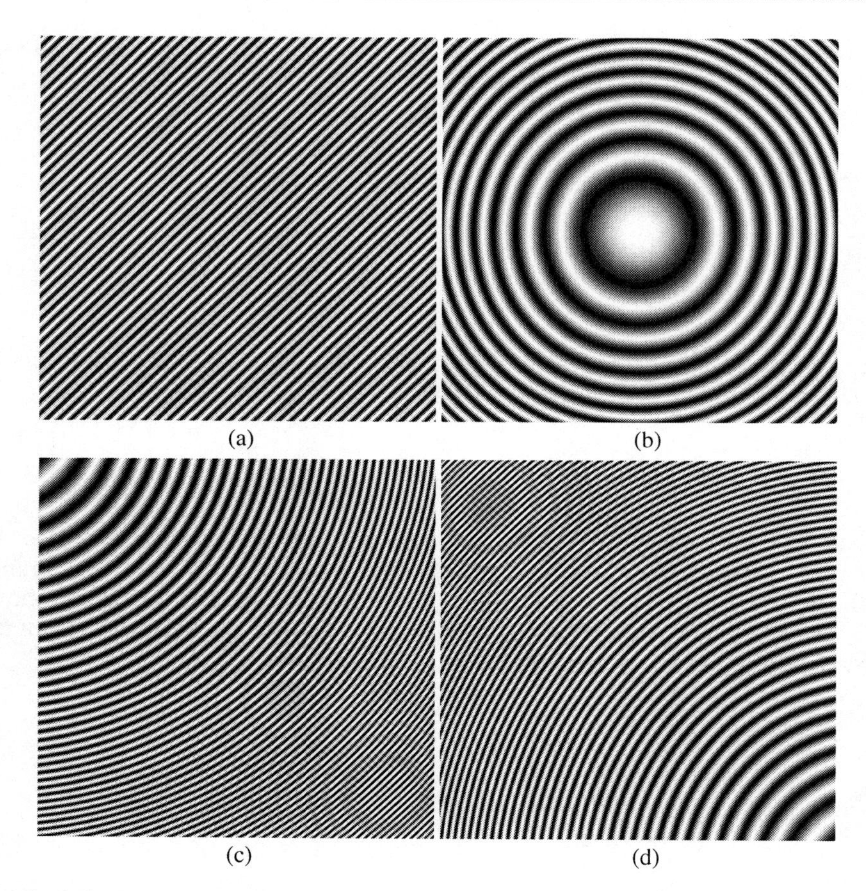

Figure 3.6 Fringe patterns. (a) Fringe pattern of the linear term; (b) Fringe pattern of the spherical term; (c) Fringe pattern of diverging spherical wavefront interfere with plane wavefront; (d) Fringe pattern of converging spherical wavefront interfere with plane wavefront

In order to further explore the interference between a spherical wavefront and a plane wavefront, we investigate the interference pattern directly from the superposition of the wavefronts. The interference between a converging spherical wavefront and a plane wavefront propagating along the optical axis is shown in Figure 3.7. The interference between a converging spherical wavefront propagating along the optical axis and a plane wavefront propagating perpendicular to the optical axis is shown in Figure 3.8. Figures 3.7c and 3.8c are the interference patterns. The comparison of the fringe patterns indicates the difference between the plane wavefronts.

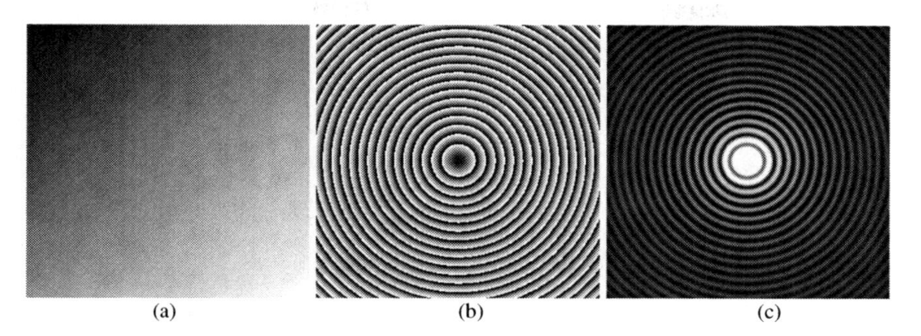

Figure 3.7 Plane wavefront and spherical wavefront interference. (a) Phase of the plane wavefront normal to the interference plane; (b) Phase of the converging spherical wavefront modulo by 2π; (c) Interference fringe

The interference between a plane wavefront propagating along the direction (cos 45°, cos 45°) and a diverging spherical wave propagating along the optical axis is shown in Figure 3.9. The interference between a plane wavefront propagating along the direction (cos 45°, cos 45°) and a converging spherical wave propagating along the optical axis is shown in Figure 3.10. Figures 3.9c and 3.10c are the interference patterns. The comparison of the fringe patterns indicates the difference between the spherical wavefronts. It is similar to the mathematical indication in Figure 3.6.

In all the above interference patterns, the spherical wavefronts are propagating along the optical axis. Thus there is an obvious amplitude difference across the fringe pattern. If the spherical wave does not propagate along the optical axis, its phase is shown in Figure 3.11b. It interferes with the same plane wave as in Figure 3.10a. The interference pattern seems like the practical hologram pattern from the digital recording device, as shown in Figure 3.11c.

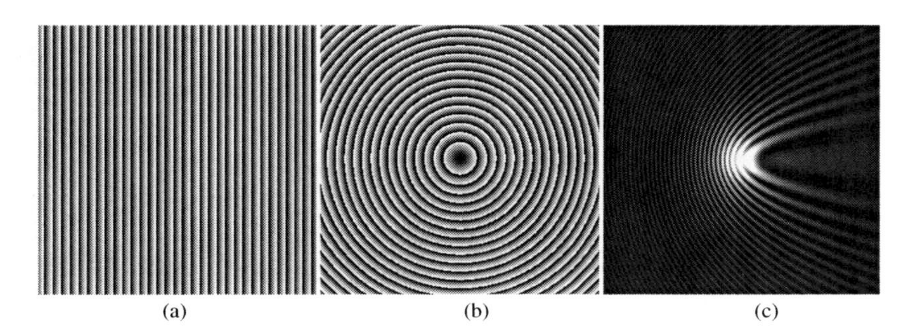

Figure 3.8 Plane wavefront and spherical wavefront interference. (a) Phase of the plane wavefront normal to the interference plane; (b) Phase of the converging spherical wavefront modulo by 2π; (c) Interference fringe

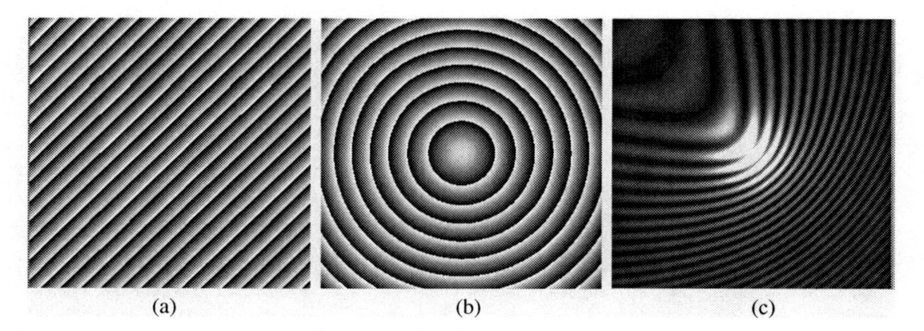

Figure 3.9 Plane wavefront and spherical wavefront interference. (a) Phase of the plane wavefront at angle 45° to the interference plane; (b) Phase of the diverging spherical wavefront modulo by 2π; (c) Interference fringe

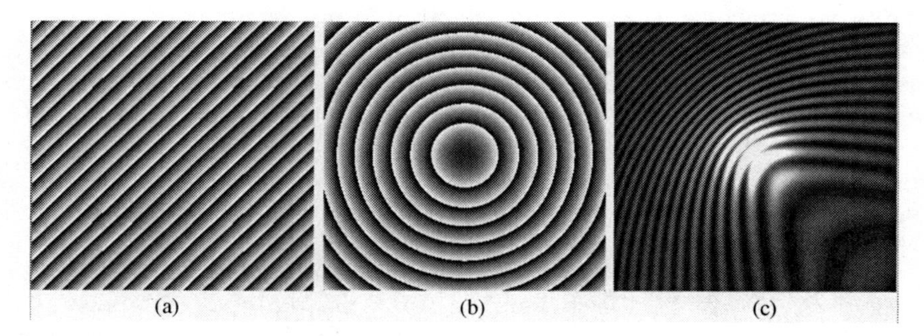

Figure 3.10 Plane wavefront and spherical wavefront interference. (a) Phase of the plane wavefront normal to the interference plane; (b) Phase of the converging spherical wavefront modulo by 2π; (c) Interference fringe

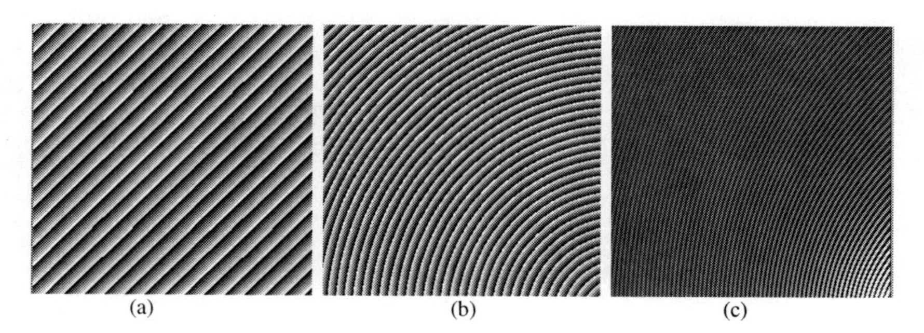

Figure 3.11 Plane wavefront and spherical wavefront interference. (a) Phase of the plane wavefront normal to the interference plane; (b) Phase of the converging spherical wavefront modulo by 2π; (c) Interference fringe

3.2.1.3 Two Spherical Waves Interference

A diverging object spherical wavefront from the point source $\left(S_{Ox}, S_{Oy}, \left(z_O^2 - S_{Ox}^2 - S_{Oy}^2\right)^{1/2}\right)$ is digitized as:

$$O_d(m\Delta x, n\Delta y) = A_O \exp\left[-j\frac{\pi}{\lambda z_O}\left((m\Delta x - S_{Ox})^2 + (n\Delta y - S_{Oy})^2\right)\right] \quad (3.18a)$$

A diverging reference spherical wavefront from the point source $\left(S_{Rx}, S_{Ry}, \left(z_R^2 - S_{Rx}^2 - S_{Ry}^2\right)^{1/2}\right)$ is digitized as:

$$R_d(m\Delta x, n\Delta y) = A_R \exp\left[-j\frac{\pi}{\lambda z_R}\left((m\Delta x - S_{Rx})^2 + (n\Delta y - S_{Ry})^2\right)\right] \quad (3.18b)$$

where A_O is the unit amplitude of the object wave; A_R is the unit amplitude of the reference wave; z_O is the distance between the object point source and the observation plane; z_R is the distance between the reference point source and the observation plane. The hologram is observed in a plane which is normal to the optical axis. The intensity is:

$$\begin{aligned} I_H(x,y) &= |O_d + R_d|^2 \\ &= A_O^2 + A_R^2 + A_O A_R \cos\left\{\frac{\pi}{\lambda}\left[\frac{\left((m\Delta x - S_{Ox})^2 + (n\Delta y - S_{Oy})^2\right)}{z_O}\right.\right. \\ &\quad \left.\left. - \frac{\left((m\Delta x - S_{Rx})^2 + (n\Delta y - S_{Ry})^2\right)}{z_R}\right]\right\} \end{aligned} \quad (3.18c)$$

In general, if S_{Ox}, S_{Oy}, S_{Rx}, S_{Ry} are partly or total different from one another, Equation 3.18c denotes an off-axis hologram. There is a special case that $S_{Ox} = S_{Oy} = S_{Rx} = S_{Ry} = 0$ and $z_O \neq z_R$. In such a case the hologram generated by the two interfered spherical waves is an on-axis hologram.

When $S_{Ox} = S_{Oy} = S_{Rx} = S_{Ry} = 0$, Equation 3.18c can be written as:

$$I_H(x,y) = A_O^2 + A_R^2 + A_O A_R \cos\left\{\frac{\pi}{\lambda}\left[\frac{(z_R - z_O)\left((m\Delta x)^2 + (n\Delta y)^2\right)}{z_O z_R}\right]\right\} \quad (3.19)$$

The on-axis interference between two diverging spherical wavefronts propagating along the optical axis is shown in Figure 3.12. When the two spherical

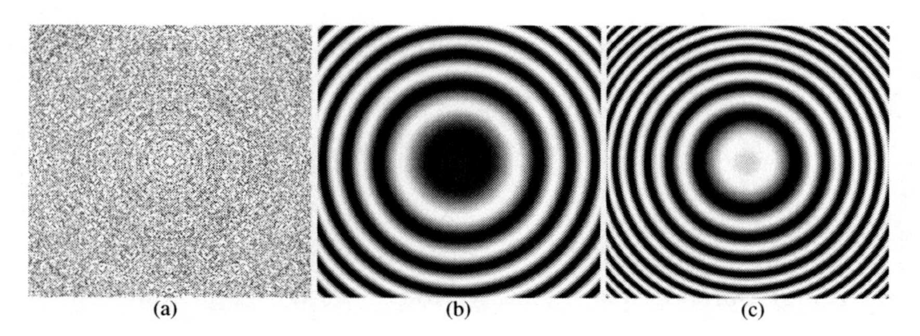

Figure 3.12 Spherical wavefronts on-axis interference. (a) $z_O = z_R$; (b) and (c) $z_O \neq z_R$

wavefronts have the same parameters (which means $z_O = z_R$), no interference will take place, as shown in Figure 3.12a. When $z_O \neq z_R$, the density of the interference fringe will increase as the increase of the difference between z_O and z_R, as shown in Figures 3.12b and c.

As concerned about the off-axis interference, the two point sources have three relative positions along the optical axis between each other which are $z_O < z_R$, $z_O = z_R$ and $z_O > z_R$, as shown in Figure 3.13. Given the certain value of S_{Ox}, S_{Oy}, S_{Rx}, S_{Ry}, the corresponding hologram fringe pattern are shown in Figure 3.14. The visible differences among the hologram fringe patterns indicate the different recording conditions of the holograms. For classical holography, the reference wavefront is the same as the recording one and is used for the optical reconstruction of the original object wavefront. Consequently the difference coming from the different recording

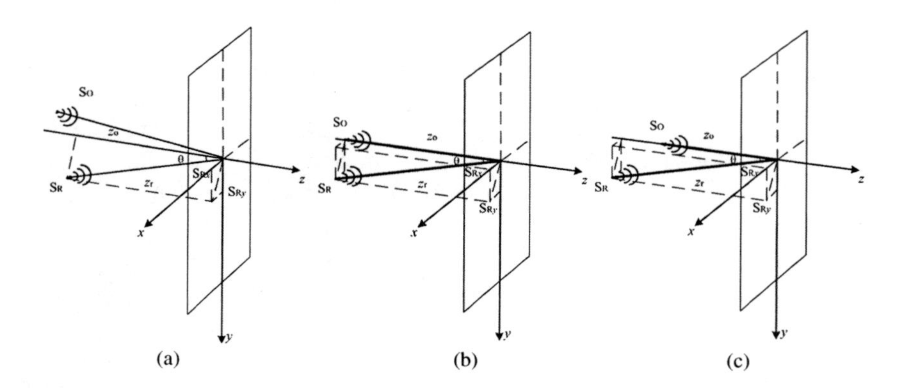

Figure 3.13 Schematic of the location of the two point sources

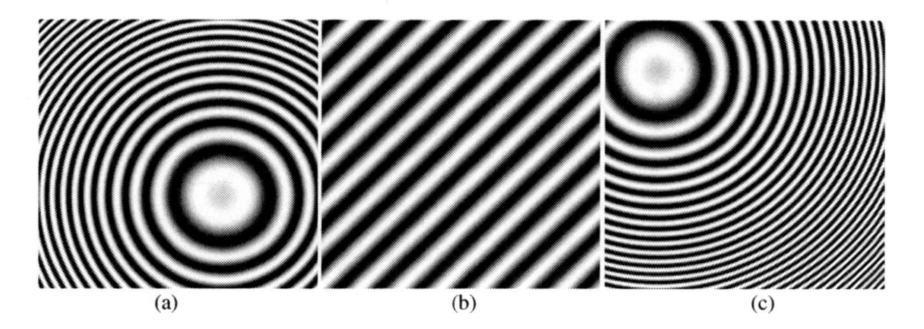

(a) (b) (c)

Figure 3.14 Off-axis hologram coming from two diverging spherical waves. (a) $z_O < z_R$; (b) $z_O = z_R$; (c) $z_O > z_R$

condition does not affect the reconstruction result. But for digital holography, it is difficult to find the same numerical reference wave as the original reference wave. The numerical reconstruction of those holograms may result in different object wavefronts by using the same numerical reference wave. Furthermore, classical holography can only be used for phase display. It cannot provide phase measurement data. For digital holography, phase data not only for the object wavefront but also for the test object can be provided. The problem of the superposition of the phase of the original object wavefront of the test object must be solved to give the correct phase measurement result. Further discussion will be given by the following hologram pattern analysis.

A converging object spherical wavefront to the point $\left(S_{Ox}, S_{Oy}, \left(z_O^2 - S_{Ox}^2 - S_{Oy}^2\right)^{1/2}\right)$ is digitized as:

$$O_c(m\Delta x, n\Delta y) = A_O \exp\left[\mathrm{j}\frac{\pi}{\lambda z_O}\left((m\Delta x - S_{Ox})^2 + \left(n\Delta y - S_{Oy}\right)^2\right)\right] \tag{3.20a}$$

A converging reference spherical wavefront to the point $\left(S_{Rx}, S_{Ry}, \left(z_R^2 - S_{Rx}^2 - S_{Ry}^2\right)^{1/2}\right)$ is digitized as:

$$R_c(m\Delta x, n\Delta y) = A_R \exp\left[\mathrm{j}\frac{\pi}{\lambda z_R}\left((m\Delta x - S_{Rx})^2 + \left(n\Delta y - S_{Ry}\right)^2\right)\right] \tag{3.20b}$$

where A_O is the unit amplitude of the object wave; A_R is the unit amplitude of the reference wave; z_O is the distance between the object point and the observation plane; and z_R is the distance between the reference point and the observation plane. The hologram is observed in a plane which is normal to the optical axis. The intensity is

$$I_H(x,y) = |O_c + R_c|^2 = A_O^2 + A_R^2 + A_O A_R \cos\left\{\frac{\pi}{\lambda}\left[\frac{\left((m\Delta x - S_{Ox})^2 + (n\Delta y - S_{Oy})^2\right)}{z_O} - \frac{\left((m\Delta x - S_{Rx})^2 + (n\Delta y - S_{Ry})^2\right)}{z_R}\right]\right\} \tag{3.20c}$$

It is obvious that Equation 3.20c has the same expression as Equation 3.18c. Consequently the fringe pattern is similar between two diverging spherical waves and two converging spherical waves.

If a diverging spherical object wave (Equation 3.18a) interferes with a converging spherical reference wave (Equation 3.20b), the intensity will be:

$$I_H(x,y) = |O_d + R_c|^2 = A_O^2 + A_R^2 + A_O A_R \cos\left\{\frac{\pi}{\lambda}\left[\frac{\left((m\Delta x - S_{Ox})^2 + (n\Delta y - S_{Oy})^2\right)}{z_O} + \frac{\left((m\Delta x - S_{Rx})^2 + (n\Delta y - S_{Ry})^2\right)}{z_R}\right]\right\} \tag{3.21}$$

Given a certain value of S_{Ox}, S_{Oy}, S_{Rx}, S_{Ry}, the hologram fringe pattern corresponding to the location of the two points is shown in Figure 3.15.

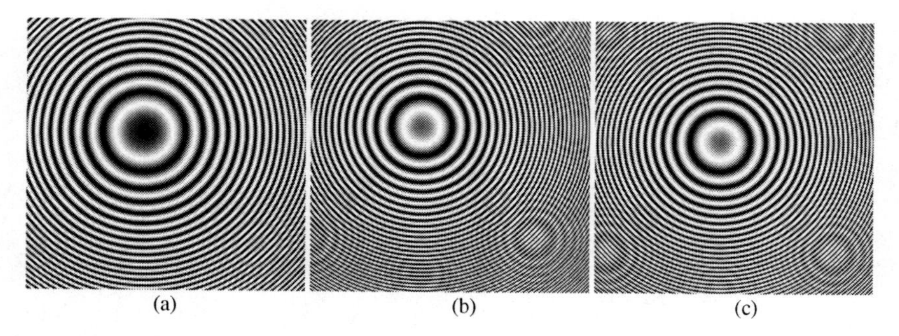

Figure 3.15 Off-axis hologram coming from a diverging spherical wave and a converging spherical wave. (a) $z_O < z_R$; (b) $z_O = z_R$; (c) $z_O > z_R$

The hologram fringe patterns are closed and unclosed circular fringes no matter what the relative positions between the two point sources. Form the hologram fringe patterns shown in Figures 3.14 and 3.15, we can conclude that the straight fringe pattern only can be achieved when the involved spherical wavefronts are of the same property.

3.2.1.4 Hologram Pattern Analysis

From the above analysis, one knows that the fringe pattern of the off-axis digital hologram can be a set of curves or straight fringes according to the involved interference phase curvature of the wavefronts, as shown in Figures 3.16a–c. The difference between Figures 3.16a and b can be directly read from the off-axis extent. But it is hard to directly tell the difference between Figures 3.16b and c due to the almost similar fringe pattern. In this case, the frequency spectra in the spatial frequency domain may provide useful information about the difference between the two digital holograms. The Fourier transform of the holograms have been taken to give the frequency spectra distribution in the spatial frequency domain as shown in

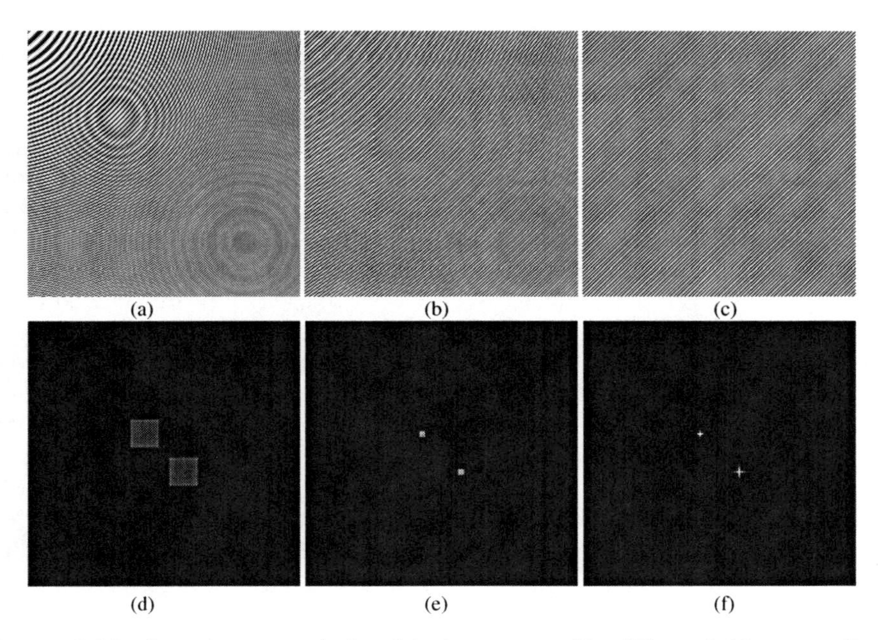

Figure 3.16 Spectrum analysis of holograms with different fringe pattern. (a) and (b) Hologram with circular fringe pattern; (c) Hologram with straight fringe pattern; (d) Fourier spectra of hologram (a); (e) Fourier spectra of hologram (b); (f) Fourier spectra of hologram (c)

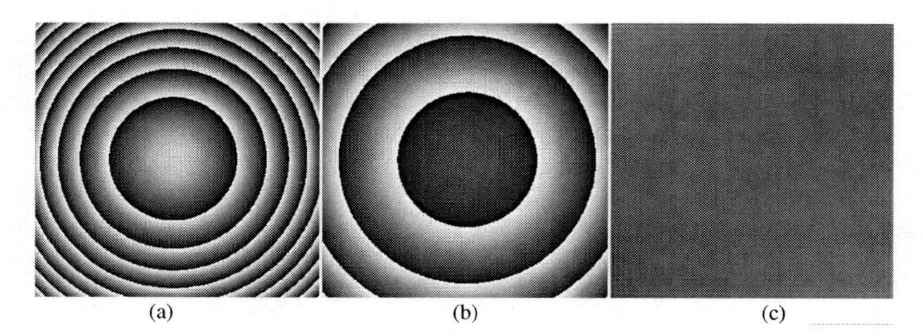

Figure 3.17 Reconstructed phase of the system with plane reconstruction reference wave normal to the hologram plane. (a) From 3.16 (a); (b) From 3.16 (b); (c) From 3.16 (c)

Figure 3.16d–f. In Figures 3.16d–f, the position of the spectrum has not changed, but the size of the spectrum has decreased from a rectangular shape to a point. From the previous analysis we know the holograms are recorded in different conditions. We wonder if such differences will affect the phase measurement of the same object wavefront in the numerical reconstruction of the digital hologram with the same numerical reference waves. As no test object is involved and the same numerical reference wave is used, we call the measured phase the system phase.

From the holograms shown in Figures 3.16a–c, the reconstruction of the system phase by using a plane reconstruction reference wave normal to the hologram plane is shown in Figure 3.17. This confirms that different recording conditions give different system phases. In Figures 3.17a–c, the spherical phase curvatures of the system phases are decreasing. In Figure 3.17c, the system phase is a constant flat. In such a case, no other phase will be introduced to the phase of the test object if it is involved. This is a very special case in the hologram recording process. It requires the same distance between the source of the illuminating wave and the hologram plane as the distance between the source of the reference beam and the hologram plane. During interference, the phase of the illuminating wave is canceled by the conjugate phase of the reference wave to give a constant flat system phase. Thus, the phase introduced by the test object is easily measured without any further numerical compensation. It is an optimized recording condition for the digital hologram given by a digital holographic microscope.

3.2.2 Digital Hologram Recording and Reconstruction

Digital holography is recording a digitized hologram by using an electronic device (e.g., a charge coupled device, CCD), and later numerically

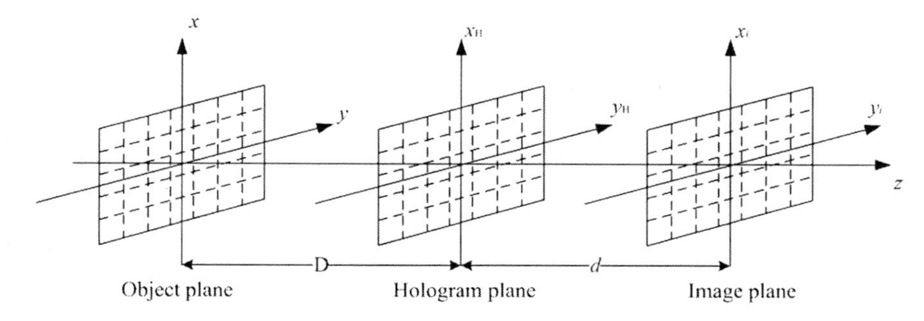

Figure 3.18 Digital hologram recording and numerical reconstruction

reconstructing with a computer both the amplitude and the phase of an optical wave arriving from a coherently illuminated object. Thus, both hardware and software are needed in digital holography. As shown in Figure 3.18, there are three planes: object plane, hologram plane and image plane. From the object plane to the hologram plane, the recording of the digital hologram is done by the optical interferometer, namely the hardware of digital holography. CCD or CMOS arrays are used to acquire the hologram and store it as a discrete digital array. From the hologram plane to the image plane, the numerical reconstruction is done by the computer software with a certain algorithm to calculate the diffraction of the waves. The digital hologram recording is described in Equation 3.13. Then digital hologram reconstruction is achieved by illumination with a numerical reference wave C. The reconstructed wavefront in the hologram plane is then given by:

$$CI_H(x,y) = C\left(|O|^2 + |R|^2\right) + CRO^* + CR^*O \tag{3.22}$$

For in-line recording geometry, the zero order and the two twin images are superposed on one another. It is hard to separate the relevant object information in one single hologram, which limits its application in real-time inspection. For off-axis recording geometry, the separation of the three terms enables a further operation on the digital hologram, such as apodization and spatial filtering [31]. After that, the terms to reproduce the original wavefront $\psi^H(x,y) = CR^*O$ in the hologram plane are achieved and propagated to the image plane to obtain a focused image $\psi^I(x,y)$.

3.2.3 Different Numerical Reconstruction Algorithms

Two different numerical reconstruction algorithms are used to calculate the scalar diffraction between ψ^H and ψ^I: the single Fresnel transform (FT) formulation [32] in the spatial domain and the angular spectrum method (ASM) in the spatial frequency domain.

Numerical reconstruction by the Fresnel transform method gives a central reconstruction formula of digital holography as follows:

$$\begin{aligned}\psi^I(n\Delta x_i, m\Delta y_i) &= e^{j\pi d\lambda\left(\frac{n^2}{N^2\Delta x_H^2}+\frac{m^2}{M^2\Delta y_H^2}\right)}\sum_{k=0}^{N-1}\sum_{l=0}^{M-1}\psi^H(k\Delta x_H, l\Delta y_H)\\ &\quad\times R^*(k\Delta x_H, l\Delta y_H)e^{\frac{j\pi}{d\lambda}(k^2\Delta x_H^2+l^2\Delta y_H^2)}e^{-2j\pi\left(\frac{kn}{N}+\frac{lm}{M}\right)}\end{aligned} \tag{3.23}$$

where m, n, k and l are $M \times N$ matrixes, they denote the address of the sampling pixel in the x and y directions respectively; d is the hologram reconstruction distance; Δx_i, Δy_i, Δx_H and Δy_H are the sampling pixel size of the image plane and the hologram plane in the x and y directions respectively; they are related by:

$$\Delta x_i = \frac{d\lambda}{N\Delta x_H}, \Delta y_i = \frac{d\lambda}{M\Delta y_H}.$$

Numerical reconstruction of ψ^H by the angular spectrum method needs a Fourier transform and an inverse Fourier transform:

$$\begin{cases}\psi^I(n\Delta x_i, m\Delta y_i) = \dfrac{\exp(jkd)}{j\lambda d}\mathrm{FFT}^{-1}\{\mathrm{FFT}\{\psi^H(k\Delta x_H, l\Delta y_H)\}\cdot G(k\Delta\xi, l\Delta\eta)\}\\ G(k\Delta\xi, l\Delta\eta) = \exp\left[j\dfrac{2\pi d}{\lambda}\sqrt{1-(\lambda k\Delta\xi)^2-(\lambda l\Delta\eta)^2}\right]\end{cases} \tag{3.24}$$

where $G(k\Delta\xi, l\Delta\eta)$ is the optical transfer function in the spatial frequency domain; $\Delta\xi$ and $\Delta\eta$ are the sampling intervals in the spatial frequency domain. The relation between the sampling intervals of the hologram plane and that of the image plane is $\Delta x_i = \frac{1}{N\Delta\xi} = \Delta x_\mathrm{H}$ and $\Delta y_i = \frac{1}{M\Delta\eta} = \Delta y_\mathrm{H}$.

Using the single Fresnel transform formulation, the resolution of the reconstructed optical wave field not only depends on the wavelength of the illuminating light but also the reconstruction distance. And it is always lower than the resolution of the CCD camera. When using the angular spectrum method, one can obtain the reconstructed optical wave field with the maximum resolution the same as the pixel size of the CCD camera. Nevertheless, both the single Fresnel transform formulation and the angular spectrum method can give a correct reconstruction of the recorded hologram. The difference between the two methods is that the single Fresnel transform formulation works in the spatial domain, while the angular spectrum method works in the spatial frequency domain. As the pre-processing of the digital hologram by apodization and spatial filtering involves its Fourier spectra, the angular spectrum method will provide a more time-saving numerical reconstruction of the recorded digital hologram.

3.3 Digital Holographic Microscopy System

The quantitative phase measurement of transparent specimens, such as living cells, micro optics, diffractive elements, and so on, is attractive because it contains a lot of information such as the shape, refractive index and other characteristics of the material. In-line phase-shifting techniques [16] and the off-axis holographic interferometry method [33] are successful means of phase measurements in digital holography. However, both need more than one digital hologram in the processing procedure. The absolute phase measurements from only one single digital hologram [22] make digital holography a real-time monitoring and inspection method.

The limited sampling capacity of the electronic camera is the reason one must find different approaches to achieve microscopic imaging with digital holography, and so revives digital holographic microscopy (DHM) [35]. DHM, through the introduction of a microscope objective (MO), gives very satisfactory measurement results both in lateral resolution and vertical resolution. Nevertheless, the MO affects the divergence of the object wave and results in a wavefront aberration between the object and reference waves. The wavefront aberration will render the phase measurement a failure if there is no powerful numerical phase compensation procedure [36, 37].

Another way to achieve microscopic imaging with digital holography is by using a lens-less DHM geometry. Diverging spherical waves are used for the hologram recording, and allow a geometrically magnified digital hologram to be numerically reconstructed with an improved resolution, without using any image-forming lens. But there is still a spherical phase curvature arising from the illuminating waves in the object phase that needs to be numerically compensated.

Numerical phase compensation is based on the computation of a phase mask to be multiplied by the recorded hologram or by the reconstructed wave field in the image plane. In the early days, the phase mask was computed depending on the precisely measured parameters of the optical set-up. It is then multiplied by the reconstructed wavefront. The correct phase map was obtained by a time-consuming digital iterative adjustment of these parameters to remove the wavefront aberration [38–40]. To exclude the necessity of physically measuring the optical set-up parameters, a phase mask is computed in a portion of the hologram where the specimen is flat and used for the aberration compensation. It was proposed by Ferraro *et al.* [41], and then developed by Colomb *et al.* [42–44] by using a polynomial fitting procedure to adjust the parameters for one dimension at first and then for two dimensions. Numerical compensation, however, makes the reconstruction algorithm complex by the iterative adjustment and extrapolation of the fitted polynomials in different areas. For the special case of the microlens shape (spherical shape), this may result in false compensation. In the case of the compensation in the reconstruction plane, the phase mask has to be adapted when the reconstruction distance is changed. Another way to remove the phase curvature is by using a reference hologram recorded by the same set-up without the test specimen [45]. Both the above methods are done numerically.

One can also physically solve the problem by introducing the same curvature in the reference wave, such as in the Linnik interferometer [27] and the Mach–Zehnder configuration [28, 29]. In these configurations, the use of the measurement optics in the reference arm duplicates the objective measurement optics in the measurement arm. The curvature of the object wave is compensated for by the reference wave during interference. It is also possible to introduce a lens in the reference arm to minimize the difference between the curvatures of the object and the reference waves [43, 46, 47]. This requires precise alignment of all the optical elements involved. However, the process may be simplified with the availability of software which can monitor the system phase in real time.

3.3.1 Digital Holographic Microscopy with Physical Spherical Phase Compensation

A digital hologram is physically recorded and numerically reconstructed. According to Equation 3.22, the phase of the term of interest is

$$\varphi(CR^*O) = \varphi(C) + \varphi(R^*) + \varphi(S) + \varphi(O_t) + \varphi(MO) \tag{3.25}$$

where $\varphi(C)$ is the phase of the numerical reconstruction reference wave front; $\varphi(R^*)$ is the phase of the conjugate recording reference wave front; $\varphi(S)$ is the phase of the illumination wave; $\varphi(O_t)$ is the phase introduce by the test specimen; and $\varphi(MO)$ is the phase coming from the imaging lens. In classical holography, the original recording reference wave is preferred for the reconstruction reference wave, which means $\varphi(C) = -\varphi(R^*)$. But in digital holography it is difficult to find the replica of the original recording reference wave. We define a system phase as

$$\varphi_{SP} = \varphi(C) + \varphi(R^*) + (\varphi(S) + \varphi(MO)) \tag{3.26}$$

This system phase must be removed from the phase that is directly reconstructed from the hologram. An easy way to do so is to use a numerical plane reference reconstruction wave which means $\varphi(C) = 0$. In other words, no additional phase will be introduced to the phase recorded by the hologram. At this point the system phase is the same as what was mentioned in Section 3.2.1.4. It includes the phase contribution of the conjugate reference wave and the illumination wave and the MO used for imaging. Fortunately, regardless of what kind of illumination wave used, it will become a spherical wave after the imaging lens [32]. It means that $\varphi(S) + \varphi(MO)$ is a spherical phase. As regards the physical compensation of this spherical phase, we can use a spherical wave as the reference wave in the recording process. There is one and only one position for the reference source point, where one can obtain $\varphi(R^*) = -(\varphi(S) + \varphi(MO))$ in the digital hologram recording process.

In the previous sections we have analyzed the interference pattern of two spherical wavefronts. For DHM, the diverging spherical wavefront should be taken into account. Based on the previous analysis, we now present a more specific theoretical analysis of the involved spherical wavefronts interference in DHM.

We assumed that the reference wave is generated by a point source located at coordinates $\left(S_{Rx}, S_{Ry}, \left(z_R^2 - S_{Rx}^2 - S_{Ry}^2\right)^{1/2}\right)$, the illuminating wave is generated by a point source located at coordinates $\left(S_{Ox}, S_{Oy}, \left(z_O^2 - S_{Ox}^2 - S_{Oy}^2\right)^{1/2}\right)$. z_R and z_O are, respectively, the distance between the source points of the reference and illuminating waves and the hologram plane. Using quadratic phase approximations to the spherical waves involved, the reference wavefront in the hologram plane is thus given by:

$$R(x, y) = \exp\left\{-j\frac{\pi}{\lambda z_R}\left[(x - S_{Rx})^2 + (y - S_{Ry})^2\right]\right\} \tag{3.27a}$$

The illuminating wave is modulated by the phase of the object. In the hologram plane, it is given by:

$$O(x, y) = A_O \exp\left\{-j\frac{\pi}{\lambda z_O}\left[(x - S_{Ox})^2 + (y - S_{Oy})^2\right]\right\}\exp[j\varphi(x, y)] \tag{3.27b}$$

where A_O is the unit amplitude and $\varphi(x, y)$ is the phase introduced by the test object. The corresponding intensity distribution in the pattern of the interference between the two waves is:

$$\begin{aligned}
I_H(x,y) = {} & 1 + |A_O|^2 \\
& + A_O \exp\left[-\mathrm{j}\frac{\pi}{\lambda}\left(\frac{S_{Rx}^2}{z_R} - \frac{S_{Ox}^2}{z_O} + \frac{S_{Ry}^2}{z_R} - \frac{S_{Oy}^2}{z_O}\right)\right] \\
& \times \exp\left[-\mathrm{j}\frac{\pi}{\lambda}\left(\frac{1}{z_R} - \frac{1}{z_O}\right)(x^2+y^2) + \mathrm{j}\frac{2\pi}{\lambda}\left(\frac{S_{Rx}}{z_R} - \frac{S_{Ox}}{z_O}\right)\right. \\
& \left. x + \mathrm{j}\frac{2\pi}{\lambda}\left(\frac{S_{Ry}}{z_R} - \frac{S_{oy}}{z_O}\right) y\right] \exp[-\mathrm{j}\varphi(\mathrm{x},\mathrm{y})] \\
& + A_O \exp\left[\mathrm{j}\frac{\pi}{\lambda}\left(\frac{S_{Rx}^2}{z_R} - \frac{S_{Ox}^2}{z_O} + \frac{S_{Ry}^2}{z_R} - \frac{S_{Oy}^2}{z_O}\right)\right] \\
& \times \exp\left[\mathrm{j}\frac{\pi}{\lambda}\left(\frac{1}{z_R} - \frac{1}{z_O}\right)(x^2+y^2) - \mathrm{j}\frac{2\pi}{\lambda}\left(\frac{S_{Rx}}{z_R} - \frac{S_{Ox}}{z_O}\right)\right. \\
& \left. x - \mathrm{j}\frac{2\pi}{\lambda}\left(\frac{S_{Ry}}{z_R} - \frac{S_{oy}}{z_O}\right) y\right] \exp[\mathrm{j}\varphi(\mathrm{x},\mathrm{y})]
\end{aligned} \tag{3.27c}$$

The interference term of interest includes combinations of a spherical wavefront, tilt in the x direction, tilt in the y direction and a constant phase. One may not be able to clearly discern them directly from the interference pattern. Its Fourier transform gives the Fourier spectra distribution

$$\begin{aligned}
I_H^F\left(f_x, f_y\right) = {} & \delta\left(f_x, f_y\right) \\
& + \mathrm{j}\lambda \frac{z_R z_O}{z_O - z_R} \exp\left[\mathrm{j}\pi\lambda \frac{z_R z_O}{z_O - z_R}\left(f_x^2 + f_y^2\right)\right] \\
& \otimes \delta\left(f_x - \frac{1}{\lambda}\left(\frac{S_{Rx}}{z_R} - \frac{S_{Ox}}{z_O}\right), f_y - \frac{1}{\lambda}\left(\frac{S_{Ry}}{z_R} - \frac{S_{Oy}}{z_O}\right)\right) \otimes \mathrm{FFT}\{\exp[\mathrm{j}\varphi(x,y)]\} \\
& + \mathrm{j}\lambda \frac{z_R z_O}{z_O - z_R} \exp\left[-\mathrm{j}\pi\lambda \frac{z_R z_O}{z_O - z_R}\left(f_x^2 + f_y^2\right)\right] \\
& \otimes \delta\left(f_x + \frac{1}{\lambda}\left(\frac{S_{Rx}}{z_R} - \frac{S_{Ox}}{z_O}\right), f_y + \frac{1}{\lambda}\left(\frac{S_{Ry}}{z_R} - \frac{S_{Oy}}{z_O}\right)\right) \otimes \mathrm{FFT}\{\exp[-\mathrm{j}\varphi(x,y)]\}
\end{aligned} \tag{3.28}$$

where $\otimes$ denotes the convolution operation. It is obvious that the spectrum of the interference term of interest (the virtual original object) consists of three parts: $j\lambda \frac{z_R z_O}{z_O - z_R} \exp\left[j\pi\lambda \frac{z_R z_O}{z_O - z_R}\left(f_x^2 + f_y^2\right)\right]$, $\delta\left(f_x - \frac{1}{\lambda}\left(\frac{S_{Rx}}{z_R} - \frac{S_{Ox}}{z_O}\right), f_y - \frac{1}{\lambda}\left(\frac{S_{Ry}}{z_R} - \frac{S_{Oy}}{z_O}\right)\right)$ and $\mathrm{FFT}\left\{\exp\left[j\varphi(x,y)\right]\right\}$. These three parts determine the shape of the spectrum. The first term is a spherical factor, which results in the spherically spectrum extension. The second term is a delta function, indicating the position of the spectrum as $\left[\left(\frac{S_{Rx}}{z_R} - \frac{S_{Ox}}{z_O}\right), \left(\frac{S_{Ry}}{z_R} - \frac{S_{Oy}}{z_O}\right)\right]$. The third term is the data about the test object.

The difference between z_R and z_O results in a different hologram pattern, thus different frequency spectra distribution in the hologram frequency domain. If $z_R > z_O$, hence, $\frac{z_R z_O}{z_O - z_R} < 0$. The spherical wavefront of the interference term of interest is a converging one. This means the divergence of the spherical wavefront coming from the illuminating wavefront or the imaging MO is smaller than that of the reference wavefront. In the interference term, the conjugate of the reference wavefront is a converging one. Consequently, a converging wavefront is left in the DHM system. If $z_R = z_O = z$, hence, $\frac{1}{z_R} - \frac{1}{z_O} = 0$. The spherical wavefront disappears and only leaves the tilt and constant phase. The pattern of the hologram is a set of straight fringes which is described by the following equation:

$$
\begin{aligned}
I_H(x,y) &= 1 + |A_O|^2 \\
&\quad + A_O \exp\left[-j\frac{\pi}{\lambda}\left(\frac{S_{Rx}^2 - S_{Ox}^2 + S_{Ry}^2 - S_{Oy}^2}{z}\right)\right] \\
&\quad \times \exp\left[j\frac{2\pi}{\lambda}\left(\frac{S_{Rx} - S_{Ox}}{z}\right)x + j\frac{2\pi}{\lambda}\left(\frac{S_{Ry} - S_{Oy}}{z}\right)y\right]\exp\left[-j\varphi(\mathrm{x},\mathrm{y})\right] \\
&\quad + A_O \exp\left[j\frac{\pi}{\lambda}\left(\frac{S_{Rx}^2 - S_{Ox}^2 + S_{Ry}^2 - S_{Oy}^2}{z}\right)\right] \\
&\quad \times \exp\left[-j\frac{2\pi}{\lambda}\left(\frac{S_{Rx} - S_{Ox}}{z}\right)y - j\frac{2\pi}{\lambda}\left(\frac{S_{Ry} - S_{Oy}}{z}\right)y\right]\exp\left[j\varphi(\mathrm{x},\mathrm{y})\right]
\end{aligned}
\tag{3.29}
$$

This means the spherical wavefront coming out from the illuminating wavefront or the imaging MO is totally compensated by the reference wavefront during interference. Consequently, a plane wavefront should be left in the

DHM system. Its Fourier transform gives the Fourier spectra distribution as follows:

$$I_H^F\left(f_x, f_y\right) = \delta\left(f_x, f_y\right)$$
$$+\delta\left(f_x - \frac{S_{Rx} - S_{Ox}}{\lambda z}, f_y - \frac{S_{Ry} - S_{Oy}}{\lambda z}\right) \otimes \mathrm{FFT}\{\exp[j\varphi(x,y)]\} \quad (3.30)$$
$$+\delta\left(f_x + \frac{S_{Rx} - S_{Ox}}{\lambda z}, f_y + \frac{S_{Ry} - S_{Oy}}{\lambda z}\right) \otimes \mathrm{FFT}\{\exp[-j\varphi(x,y)]\}$$

If there is no test object, the spectrum of interest will be a delta function with sharp point distribution. When a different object is being tested, the convolution in Equation 3.30 will make the sharp point distribution a complicated shape.

If $z_R < z_O$, hence, $\frac{z_R z_O}{z_O - z_R} > 0$. The spherical wavefront of the interference term of interest is a diverging one. This means the divergence of the spherical wavefront coming from the illuminating wavefront or the imaging MO is bigger than that of the reference wavefront. Consequently, a diverging wavefront is left in the DHM system.

In conclusion, when $z_R \neq z_O$, the left spherical wavefront can be either a diverging one or a converging one depending on the relative position of the two point sources. This means the spherical wavefront coming from the illuminating wavefront or the imaging MO cannot be physically compensated by the reference wavefront during interference. When an individual interference term of interest is considered, a system phase with a spherical curvature is presented. For a numerical reconstruction, a collimated reference wave is used in the off-axis digital holographic microscopy set-up [38–41, 48], due to the simplicity of the digital replica of such a reference wave. Thus, no other spherical phase will be introduced to the whole interference term of interest. For the desired phase of the test object, the system spherical phase must be numerically compensated.

When $z_R = z_O$, the spherical wavefront can be removed by the physical matching of the involved object and reference wavefront. Thus the phase directly reconstructed from the hologram is the phase introduced by the test object without any other further numerical process.

From the analysis above, it is obvious that the shape of the spectrum can indicate whether the wavefront aberration between the object and reference waves can be physically compensated during hologram recording. The numerical reference wavefront should be carefully chosen to ensure no other phase factor is introduced into the reconstructed object phase. One can monitor the shape of the spectrum to judge whether the spherical phase curvature is totally compensated in the set-up alignment process.

3.3.2 Lens-Less Common-Path Digital Holographic Microscope (49)

In this section, we propose a transmission DHM system based on a single cube beam splitter (SCBS) interferometer [50], using a diverging spherical wave to provide magnification of the test specimen. This is a lens-less imaging geometry. The use of the beam splitter cube makes it a symmetrical common-path interferometer. This means the object wave and the reference wave travel the same distance to reach the hologram plane. Hence, the spherical phase curvature introduced by the illuminating wave is physically compensated by the reference wavefront during interference according to Equation 3.29.

The DHM system is shown in Figure 3.19. Light emitting from a laser diode is shaped and filtered by a small pinhole (1 μm in diameter) to serve as the point source. Diverging spherical waves from this point source will ensure the necessary magnification with the lens-less geometry. The location of the point source together with the position of the test specimen and CCD camera determines the magnification of the lens-less system. The light wave transmits from the sample plate and incident on a BS with its central semi-reflecting layer placed at a small angle along the light propagation direction. The test sample is located in the top half of the beam path which is reflected by the central semi-reflection layer of the BS and acts as the object arm of the interferometer. The bottom half transmits through the central semi-reflecting layer and acts as the reference arm. The rays can be traced from the sample by the dashed lines. A CCD camera records the interference pattern at one side of the exact output plane of the SCBS.

The ray trajectories in the SCBS are shown in Figure 3.20. Light incident on the front cube wall will change its propagation direction inside the SCBS. Only the refracted light which arrives at the central semi-reflecting layer and is

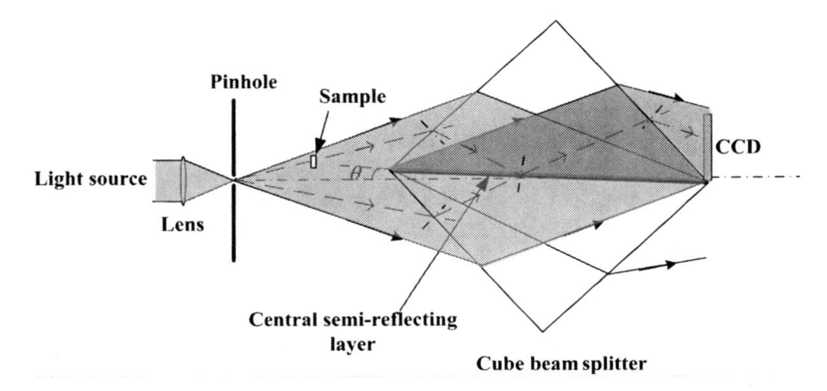

Figure 3.19 DHM system based on a SCBS interferometer
Note: The dashed lines are used to illustrate the ray tracing from the sample

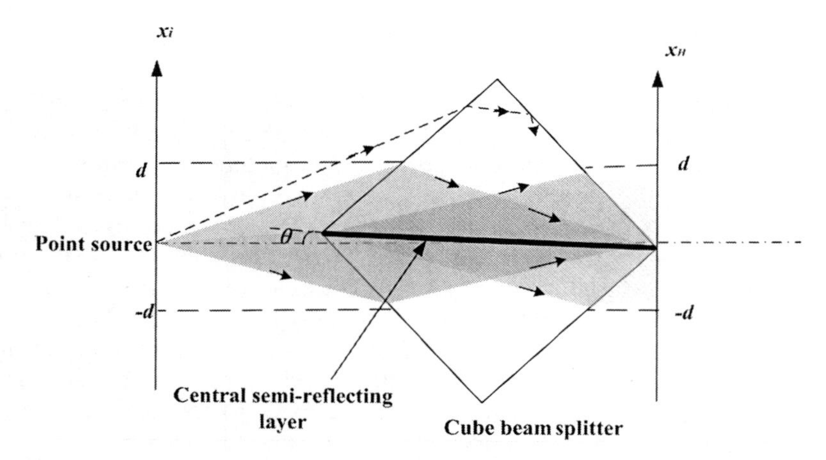

Figure 3.20 Ray trajectories in the beam splitter cube
Note: d is the size (in the x-direction) of the working region. More external rays suffer total reflection at the cube walls

reflected to the back cube wall can be output through the BS. Other rays will travel from the front cube wall directly to the back cube wall and suffer total reflection. This can be demonstrated by an experiment where the working region in the x-direction is the size of the order of 30% of the cube's lateral size and in the y-direction, the size of the order of 100% of the cube's lateral size. Because of the small angle θ between the light propagation direction and the central semi-reflecting layer, a wedge-shaped optical path difference will be introduced to the reflected light and the transmitted light. Consequently, in the interference plane, one can achieve an interferogram with straight fringe patterns.

As the angle θ increases, higher space frequency fringe patterns can be observed as shown in Figures 3.21a and b. The smaller the space frequency of the fringe pattern, the larger the separation between the three orders of the spectrum, as shown in Figures 3.21c and d. Usually, the separation between the first order and the zero order is preferred to be as large as possible to ensure the selection of the spectrum with the largest filter. Thus, it gives an off-axis digital hologram which is more suitable for numerical reconstruction.

Since the object and reference beams are traveling along a common path to form the interference pattern, the DHM system can physically compensate the spherical phase curvature arising from the illuminating waves according to Equation 3.29. The constant phase left is independent of time. Therefore, the set-up is clearly insensitive to external vibrations as in the common path interferometer. This is not the case with "standard" DHM systems because

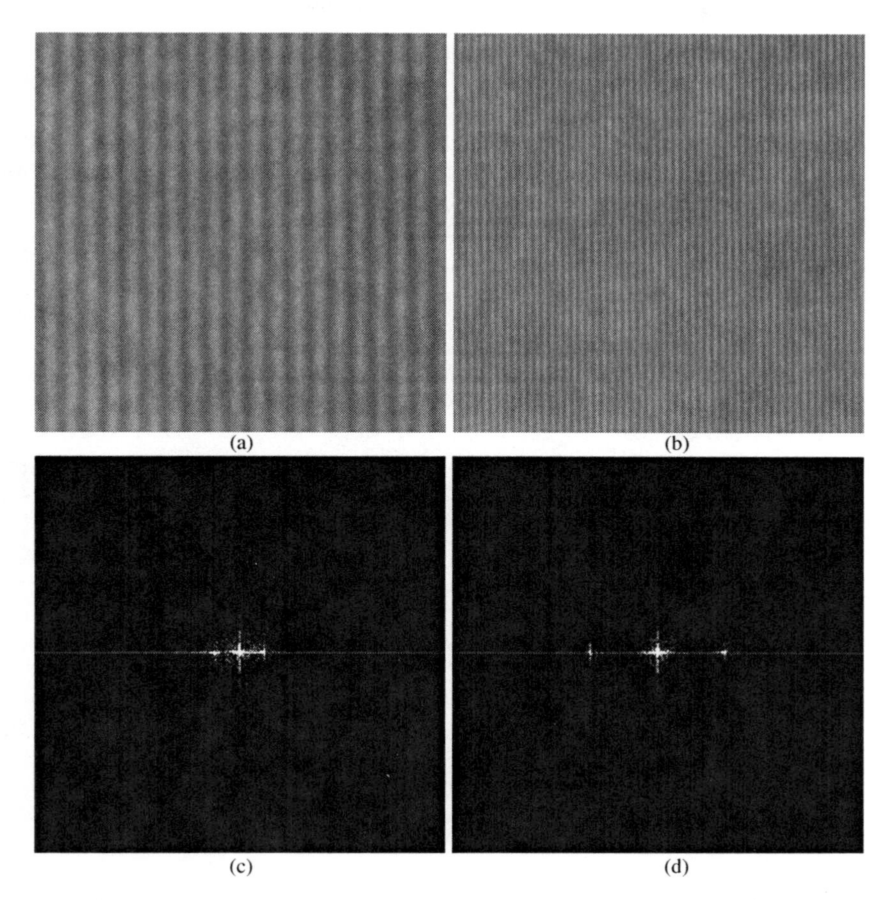

Figure 3.21 Straight fringe pattern observed for different angle θ. From (a) to (b) θ increases; (c) Fourier spectra of hologram (a); (d) Fourier spectra of hologram (b)

vibrations introduce a temporal phase offset from the different optical paths in the reference and object arms.

A plane wavefront is taken as the numerical reference wave. As the pre-processing of the digital hologram by apodization and spatial filtering involves its Fourier spectra, the angular spectrum method in the spatial frequency domain is used for the numerical hologram reconstruction.

In the spatial frequency domain, the spectrum for the original test object is filtered out and moved to the center to remove the tilt in phase.

$$\psi^{HF}\left(f_x, f_y\right) = \delta\left(f_x, f_y\right) \otimes \mathrm{FFT}\{\exp[\mathrm{j}\varphi(x, y)]\} \tag{3.31}$$

where ψ^{HF} denotes the Fourier transform of the original test object. Then, the numerical reconstruction by the angular spectrum method only needs an inverse Fourier transform, according to Equation 3.24.

In the hologram recording process, we used a spherical wavefront. In the hologram numerical reconstruction process, we used a plane wavefront. Thus, the reconstruction distance must be calculated by:

$$d = \left[\frac{1}{d_r'} - \frac{\lambda'}{\lambda}\left(\frac{1}{d_r} - \frac{1}{d_o}\right)\right]^{-1} \tag{3.32}$$

where d_r and d_r' describe the distances between the source point of a spherical reference wave and the hologram plane in the recording and reconstruction processes respectively. λ and λ' are the wavelengths for recording and reconstruction. d_o describes the distance between the object and the CCD. Here, because of the plane wavefront, $d_r' \to \infty$, the plane reference reconstruction wavefront introduces a maximum lateral magnification of the reconstructed image ψ^I with $M_{\max} = \frac{d_r}{d_r - d_o}$. This maximum lateral magnification is determined by the recording geometry of the digital hologram.

The quantitative phase of the test object can be obtained by Equation 3.33:

$$\varphi(x, y) = \arctan\frac{\mathrm{Im}\left(\psi^I(x, y)\right)}{\mathrm{Re}\left(\psi^I(x, y)\right)} \tag{3.33}$$

A sample consisting of a glass substrate with a microscopic laser-ablated spot was investigated to establish the viability of the proposed set-up. In the experiment, a 60× objective lens with a 1 μm pinhole was used to spatially filter a 407 nm laser to improve the spatial coherence of the source and obtain a higher divergence. The specimen was placed between the pinhole and the beam splitter cube (25 mm in size). The digital holograms were recorded by using a 1280 × 960 pixels CCD with 4.65 μm square pixels. The recorded hologram was reconstructed to give the intensity and quantitative phase of the object.

The digitally recorded hologram is shown in Figure 3.22. The distance between the test specimen and the point source (at the exit of the pinhole) is 40 mm. The distance between the CCD and the point source is 200 mm. So, the magnification arising from the recording geometry is 5×. The intensity and 3D phase of the microscopic laser-ablated structure on glass have been numerically reconstructed. The images agree favorably with that obtained from a standard imaging microscope as shown in Figure 3.23. The orthogonal fringes appearing in Figure 3.23a are caused by the diffraction calculation method itself. The rectangular boundary of the hologram is diffracted

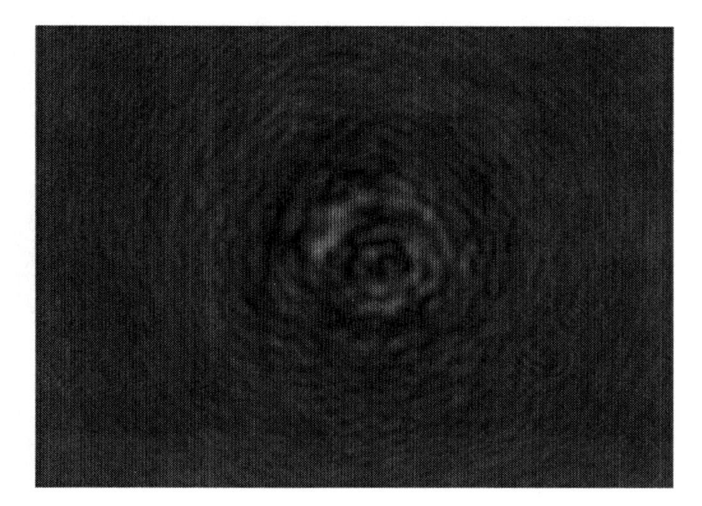

Figure 3.22 Digital hologram of a laser-ablated microscopic spot on a glass substrate

in the process of the numerical reconstruction. The 3D quantitative phase map is shown in Figure 3.24.

The use of the beam splitter in a non-conventional way will also introduce aberrations. There may be a small relative rotation of the two rectangular prisms that form the cube when cemented together so that the fringe pattern shows a horizontal component. This will make the spectrum tilt a little in the vertical direction. This tilt aberration is the major aberration introduced by the cube beam splitter. It can be compensated by the spectrum centering process, though not totally because of the same reason described above.

When the light arrives on the beam splitter, it should diffract on the front edge. But the distance between the cube beam splitter and the CCD (the

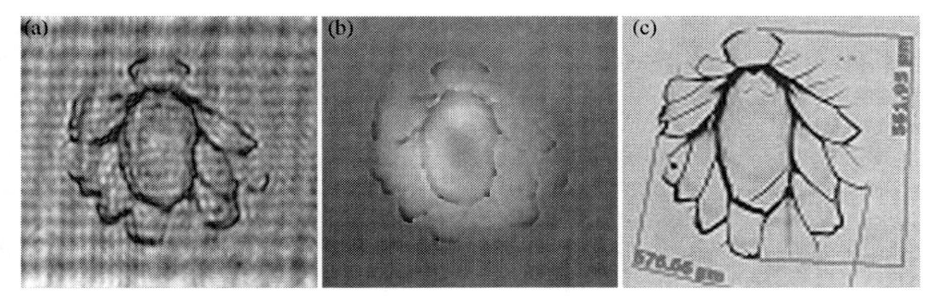

Figure 3.23 (a) Numerically reconstructed intensity image from hologram shown in Figure 3.22; (b) Phase contrast image; (c) Image from conventional microscope with 10 × objective lens

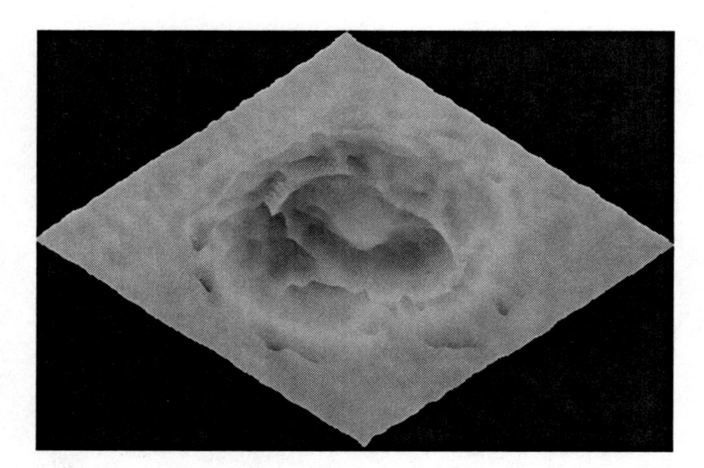

Figure 3.24 3D quantitative phase distribution of the laser mark

diagonal length of the cube beam splitter) is short enough so that the diffraction is weaker than the diffraction of the test specimen. Also, it is shown as a boundary diffraction pattern in the whole interference region. The recording of the digital hologram in the center of the whole interference region can help to suppress this effect.

The size of the beam splitter cube will limit the interception of the high frequency of the specimen. This limits the numerical aperture of the whole set-up and affects the lateral resolution of the microscope. However, the simplicity of the experimental configuration, together with a simple numerical reconstruction algorithm, makes it very attractive for DHM.

3.3.3 Common-Path Digital Holographic Microscope (51, 52)

Although the diverging wave can provide enough magnification, the performance of the CCD has a disadvantage when recording high frequency light waves. So the actual magnification that can be reached by the above DHM system is far from what is needed. In order to achieve higher magnification, we proposed another transmission DHM system based on SCBS interferometer, using an MO to provide magnification for the test specimen.

A digital off-axis hologram can be recorded by using a SCBS in a nonconventional configuration so as to both split and combine the wavefronts emerging from a microscope objective. As shown in Figure 3.25, light incident on a SCBS is split into two paths by its front edge. Each path of the light will change its propagation direction inside the SCBS. The refracted light arriving at the central semi-reflecting layer will both be reflected and transmitted to the back cube wall. The wave reflected from the left optical path will combine with the transmitting wave from the right optical path at the exit of the SCBS. Since

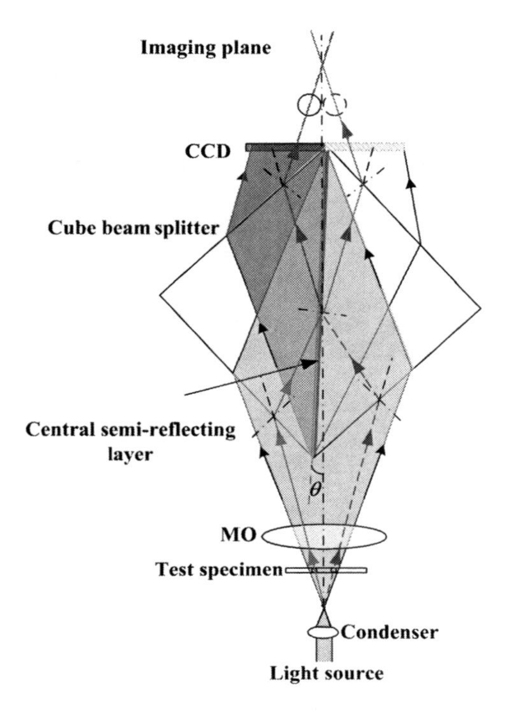

Figure 3.25 Schematic of the SCBS microscope
Note: The rays in the solid line are used to illustrate the left part of the propagating light; the rays in the dashed line are used to illustrate the right part of the propagating light

there is a small angle between the light propagation direction and the central semi-reflecting layer, a wedge-shaped optical path difference will be introduced between the reflected light and the transmitting light. Thus, two phase-shifted interferograms can be obtained at the back edge of the SCBS. Each one can be captured by the CCD camera and numerically reconstructed to give the phase of the test object. As it is a common-path interferometer, the reference arm travels the same distance as the object arm. Their spherical phase curvature is physically compensated by the other during interference in the digital hologram recording process. In the numerical reconstruction process, one needs to choose a correct plane reference wave to remove the sub-pixel off-axis tilt (if there is one) and give a flat phase surface, as shown in Figure 3.26. In this condition, the phase introduced by the test object can be easily achieved without any numerical phase compensation procedures by only one shot.

The unique imaging property of the SCBS microscope is its four-channel imaging. As shown in Figure 3.25, the light wave transmits from the sample

Figure 3.26 Phase with physical spherical phase compensation given by the SCBS microscope

plate and is incident on the SCBS with its central semi-reflecting layer placed at a small angle along the light propagation direction. The test sample located in the left half of the beam path is reflected by the central semi-reflection layer of the SCBS and captured by the left camera. At the same time, the light transmitting through the central semi-reflection layer reaches the right camera. This dual-channel imaging is indicated by the rays in the solid line in Figure 3.25. It shows that the object can be imaged on both sides of the SCBS in a plane which is perpendicular to the optical axis.

Similar to the left optical path, the right optical path also can perform the dual-channel imaging. The right half of the beam path is reflected by the central semi-reflection layer of the SCBS and at the same time is transmitting through it. The direction of the light propagation is indicated by the rays in gray. It provides a reflected image as well as a transmitting image.

The dual-channel imaging of the right half of the beam path is demonstrated in Figure 3.27. The CCD camera is put as close as it can to the back edge of the SCBS. From Figure 3.27a, one can see the edge of SCBS in the unique hologram with phase shifted interference fringe. The phase extracted from the above hologram is shown in Figure 3.27b. Its quantitative height map (in μm) is shown in Figure 3.27c which is calculated from the unwrapped phase of Figure 3.27b. This demonstration shows that the dual-channel imaging gives opposite phases of the text object due to the phase shift π in the central semi-reflection layer of the SCBS. In general, when a single lens is measured, the dual-channel imaging is not used. The CCD camera is located on the left or on the right to make sure only one channel is used. Thus, one can get the maximum view field including the object of interest.

A 1 mm × 12.8 mm × 1.2 mm refractive plano-convex linear microlens array from SUSS micro optics is tested to demonstrate the characterization of lens by the SCBS microscope. The wavelength of the light is 633 nm. A 40× microscope

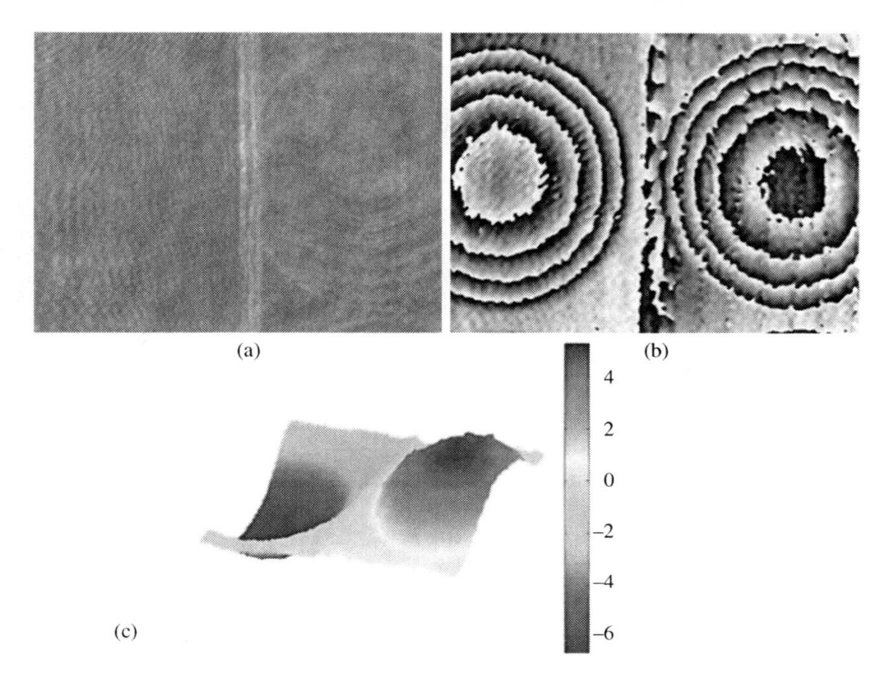

Figure 3.27 Experimental demonstration of the dual-channel imaging property of the SCBS microscope. A refractive plano-convex micro-lens is taken as the test sample. (a) Two phase-shifted interferograms captured in a single camera exposure; (b) Phase extracted from the hologram modulo by 2π; (c) Quantitative height map (in μm) which is calculated from the unwrapped phase of (b)

objective from Olympus is used for the microscope imaging. A $25 \times 25 \times 25\,\text{mm}^3$ SCBS is used in the optical path. The hologram is recorded by a 1280×960 CCD sensor with $4.65\,\mu\text{m}^2$ square pixels.

With a known refractive index of the lens material, the geometrical thickness of the lens can be deduced from the quantitative phase map, as well as the lens shape, height, and radius of curvature (ROC) if the lens has a flat face (for example, a plano-convex lens). When light is transmitted through the lens, the optical path length will change according to the height and refractive index of the lens. By using DHM, the optical path length change can easily be achieved from the phase map of the wavefront. Given the refractive index, one can calculate the height of the test lens according to the following equation:

$$h = \frac{\lambda}{2\pi} \frac{\varphi}{(n_L - n_S)} \tag{3.34}$$

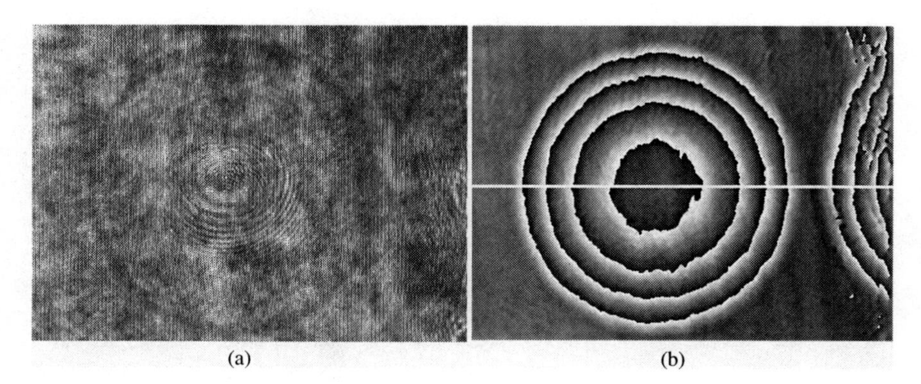

Figure 3.28 (a) Digital hologram of the lens, (b) Phase reconstructed from the hologram modulo by 2π

where λ is the wavelength of the light; φ is the phase given by the SCBS microscope; n_L is the refractive index of the lens; and n_S is the refractive index of the medium around the test lens.

The experimental captured hologram is shown in Figure 3.28a. The hologram is reconstructed by using the angular spectrum method with a normal plane reference wave at the distance 50 mm. The phase reconstructed from the hologram modulo by 2π is shown in Figure 3.28b. It should be pointed out that the reconstruction distance is not a fixed one. One can change it through the control of the focusing of the image. It means that for a different object distance, there will be a different reconstruction distance. Thus the microscope needs to be calibrated for the actual magnification it can provide. The calibration of the SCBS microscope has been done by using the 1 mm/100 division stage micrometer. The actual magnification of the system is given as 33×. Given the refractive index is 1.457 at 633 nm, as shown in Figure 3.29, the 3D quantitative height map of a single microlens is obtained from the unwrapped phase of Figure 3.28b.

The height profile of the single microlens is shown in Figure 3.30. It was drawn from the height map along with the same position as the white solid line in Figure 3.28b. Given the height profile of the lens, the radius of curvature can be calculated by the following equation:

$$ROC = \frac{h}{2} + \frac{D^2}{8h} \tag{3.35}$$

where h is the height of the microlens; D is the diameter of the microlens. The maximum height of the microlens, h, is read as 5.46 μm. The diameter, D, is read as 120 μm. Thus the calculated ROC is 332 μm. It is slightly different from the provided ROC value of 315 mm by the supplier.

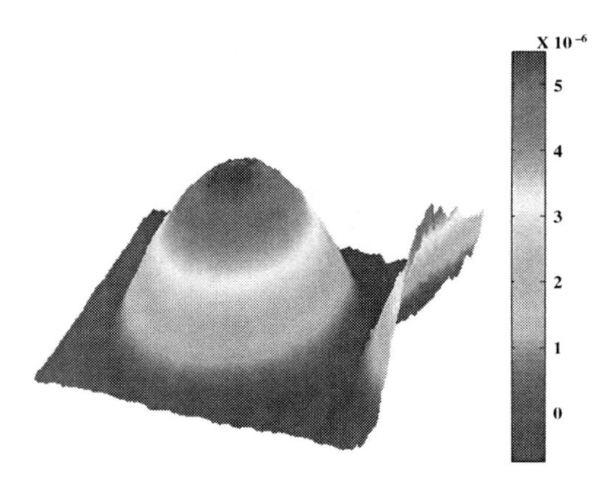

Figure 3.29 3D quantitative height map for a single refractive plano-convex microlens obtained with the SCBS microscope

If the object is not only in the left half of the beam path but also in the right half of the beam path, one can put the CCD camera either on the left or on the right edge of the beam splitter to get another dual-channel image. One image is formed by the reflection of the left part of the beam path, and the other image is formed by the transmission of the right part of the beam path. As shown in Figure 3.25, the rays in the solid line are used to illustrate the image formation for the light reflected from the central semi-reflecting layer; and the rays in the dashed line are used to illustrate the image formation for the light transmitted

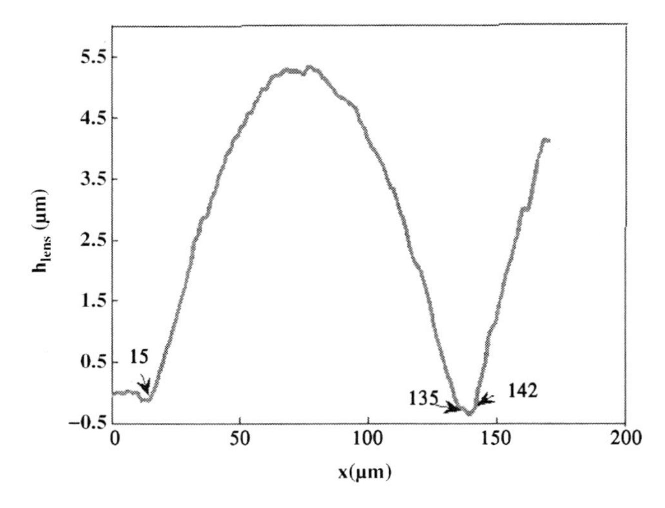

Figure 3.30 Height profile of the single refractive plano-convex microlens drawn along with the same position as the white solid line in Figure 3.28 (b)

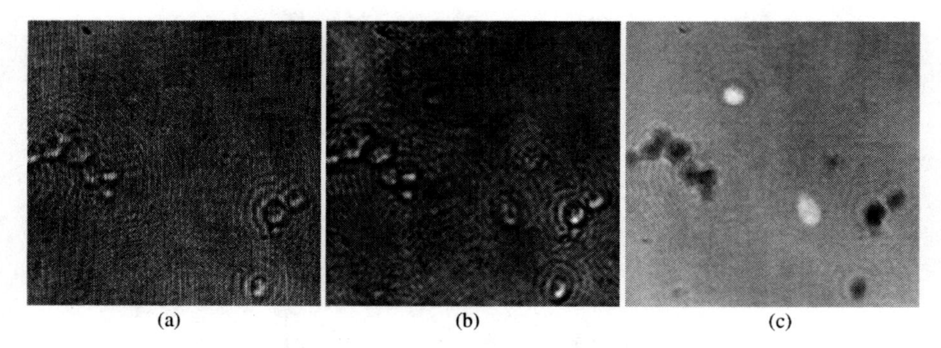

Figure 3.31 Experimental demonstration of the dual-channel imaging property of the SCBS microscope. (a) Two phase-shifted images are captured in a single hologram; (b) Intensity image extracted from the hologram; (c) Phase contrast image extracted from the hologram

through the central semi-reflecting layer. The two images will be imaged and overlapped in the CCD camera at the same time. If an object with a similar shape and material is tested, there will be a phase difference π between the two images. This unique imaging property can be applied to the comparison between the adjacent lenses in a microlens array. Thus it provides us with a way of uniformity inspecting the microlens array together with a precision sample stage.

This dual-channel imaging of the SCBS microscope is demonstrated in Figure 3.31. Yeast cells have been used as the sample for the demonstration. They are put into the c-chip with a low density. The view field is carefully chosen to make sure there is one cell in the right half of the optical path and another cell in the left half of the optical path. The CCD camera is put as close as it can to the left side of the back edge of the SCBS. The digital hologram is shown in Figure 3.31a. The hologram has been reconstructed by the angular spectrum. The reconstructed intensity and phase contrast image are shown in Figures 3.31b and c. Although in the hologram the image coming from the different channel is not very clear, one can observe them from the intensity and phase contrast image. This demonstration shows that the dual-channel imaging gives opposite phases of the text object due to the phase shift π between the reflection image and the transmitting image. If the object in the left half of the optical path is the same as the one located in the right half of the optical path, and they are located symmetrically to the optical axis, the two images will overlap. For the phases, they will be canceled out by each other.

The above dual-channel imaging can be applied to the uniformity inspection of the microlens array. Part of the test microlens array is located in the left half of the beam path which is reflected by the central semi-reflection layer of the SCBS and acts as the object arm of the interferometer. Part of the test

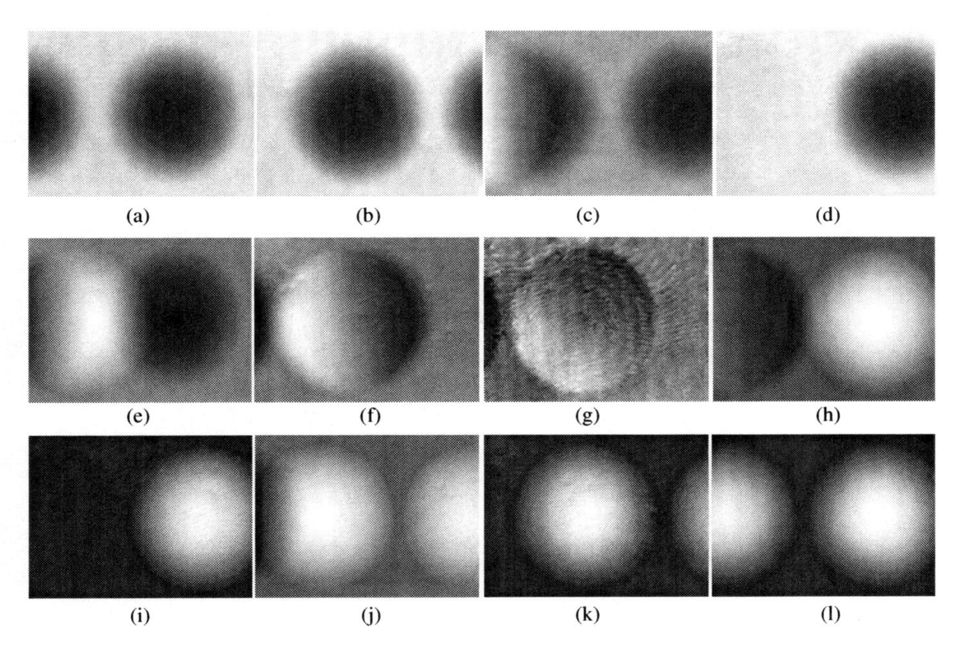

Figure 3.32 Uniformity inspection of the linear microlens array. (a) Microlens array is put in the right half beam path; (b) (d) (g) (i) (k) Symmetric microlens in the both halves of the beam path, thus compensation of the microlens can be observed; (c) (e) (f) (h) (j) Non-symmetric microlens in both halves of the beam path, thus the phase of the lens cannot be fully compensated; (l) Microlens array is moved to the left half beam path

micro-lens array is located in the right half, which transmits through the central semi-reflecting layer and acts as the reference arm. There is a phase shift π between the reflected image and the transmitting image. We can control the position of the sample and put the two images in the same place. Serious experimental results are presented in Figure 3.32 with the CCD camera put on the left side. The test sample is moved from the right to the left. At first, the sample is only present in the right half of the beam path. Consequently the phase of the lens is achieved, as shown in Figure 3.32a. As the sample moves continuously to the left half of the beam path, there will be a position where the single microlens is separated into two halves by the optical axis. One half of the microlens is imaged by the reflected light of the central semi-reflecting layer. The other half of the microlens is imaged by the light transmitting through the central semi-reflecting layer at the same position. Owing to the phase shift π between the two phases, they will compensate each other if the two halves have the same profile. In such a case, the phase will be a flat plane. As shown in Figure 3.32b, the half of the microlens in the left hand is compensated by the other half of the same microlens. Due to the limited sensor area of the CCD camera, we can only see part of the compensated microlens. The microlens array is continuously moving to the left. Two lens imaged at different positions will partly overlap. In such a case the phase cannot be fully compensated due to the different image position as shown in Figure 3.32c. In a continuous movement of the sample, full compensation and part compensation can be observed. In Figures 3.32b, d, g, i and k, the opposite phases are compensated by each other and give a flat plane due to the same shape of the lens. While in Figures 3.32c, e, f, h and j, the position of the lens is different, thus they cannot compensate each other. Finally, the sample is moved to the left half of the beam path. Consequently, as shown in Figure 3.32l, the phase of the lens is achieved opposite to the phase in Figure 3.32a. If the relative position of the lens array is under control, every lens can easily be compared with another one to give a uniformity estimation of the whole piece of the microlens array.

3.3.4 Digital Holographic Microscopy with Quasi-Physical Spherical Phase Compensation : Light with Long Coherence Length (53)

As the use of the MO only in the object beam introduces a spherical phase curvature [32], a spherical wavefront in the reference beam may provide quasi-physical spherical phase compensation. In this section, we propose transmission mode DHM systems based on the Michelson interferometric configuration, which gives quasi-physical spherical phase compensation in the recording process of the digital hologram.

One of the proposed DHM systems is built in the Michelson interferometer configuration which directly uses the light coming from the fiber point as the reference beam. As shown in Figure 3.33, light emitted from a 633nm He-Ne laser is coupled with a 2×2 single mode fiber splitter. Light from fiber point 1 together with a condenser serves as the illumination wave. A microscope objective (Mitutoyo infinity-corrected long working distance objective, $50\times$, 0.55 NA) together with a tube lens performs the imaging of the test specimen. The specimen is placed in an xyz-axis adjustable sample stage between the condenser and the MO. The imaging is obtained by adjusting the z-axis of the sample stage. The location of the MO is fixed both to the light source and the CCD camera (the hologram recording plane). Thus, the phase curvature in the object beam path will not be changed in the imaging process. In the reference beam path, the position of light source 2 can be adjusted easily in order to control the phase curvature to match that of the object wavefront when it reaches the hologram recording plane.

In order to generate interference fringes, the phases of the individual waves must be correlated in a special way. This correlation property is called coherence. Coherence is the ability of light to cause interference. The two aspects of

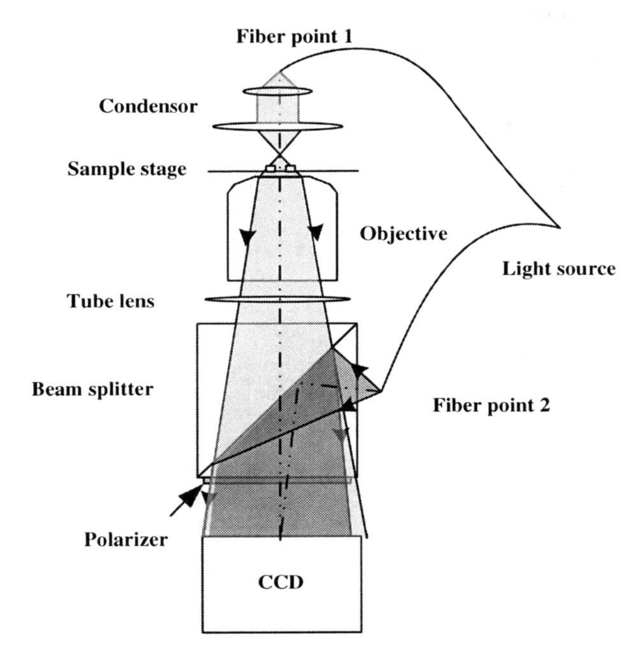

Figure 3.33 Schematic of the DHM system

coherence are the temporal and the spatial coherence. Spatial coherence depicts the mutual correlation of different parts of the same wavefront. Generally it can be improved through spatial filtering of the light source used. Temporal coherence describes the correlation of a wave with itself at different instants. Its quantitative description is its coherence length. Light with a long coherence length is called highly monochromatic. Lasers have coherence lengths from a few millimeters (for example, a multi-mode diode laser) to several hundred meters (for example, a stabilized single mode Nd:YAG laser) up to several humdred kilometers for specially stabilized gas lasers used for research purposes. Here in the Michelson-interferometer, an He-Ne laser with a coherence length of about 1 meter is used for the digital hologram recording. Consequently the optical path length difference between the object and reference path can be very large.

Regarding the physical spherical phase compensation, this is an optimized recording condition for the digital hologram. It is independent of the interferometer configuration and requires a rigorous alignment of all involved optical components. To achieve this, one needs to resort to the real-time phase monitoring function provided by the numerical reconstruction software.

The software must be designed for the real-time display of the reconstructed intensity and phase. Its key is the removal of the off-axis tilt. Different methods

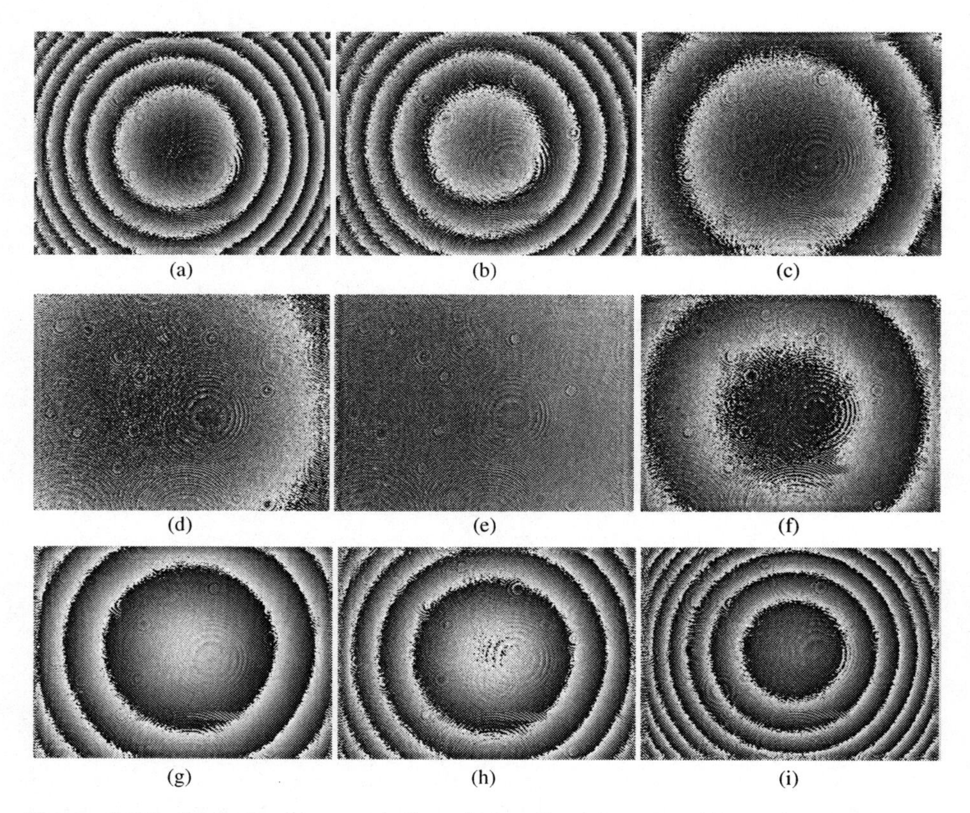

Figure 3.34 System phase variation during the change of the reference wave. (a)–(d) Converging spherical phase; (e) Quasi-flat phase; (f)–(i) Diverging spherical phase

exist to remove this tilt. One is the maximum value orientation of the spectrum of interest and its centralization in the frequency spectrum domain. This is an easy way to quickly remove the tilt but it seems useless to remove the tilt for the sub-pixel. In this case, a plane reference wave is a better choice. In our software, we take the combination of the two methods to remove the off-axis tilt. When the off-axis tilt is removed, this leaves only a system phase because of the difference between the object wave and the reference wave. As shown in Figure 3.34, the system phase changes its direction during the adjustment of the reference wave in a fixed direction towards the CCD recording plane.

When $z_R > z_O$, a converging spherical phase results from the conjugate reference wave as shown in Figures 3.34a–d. When $z_R < z_O$, a diverging spherical phase results from the object wave as shown in Figures 3.34f–i. In between the converging spherical phase and the diverging spherical phase, there is only one plane to achieve a quasi-flat phase where z_R must be exactly equal to z_O, as shown in Figure 3.34e.

It is necessary to use the real-time phase monitoring in the optical components alignment to ensure the compensation of the system spherical phase. There is a mutual interaction between the hardware (optical interferometer) and the software (numerical reconstruction). The precise alignment of all the optical components will make a sharp distribution of the interested spectrum. It in turn is used for an easy spectrum selection and centralization in the numerical reconstruction. The DHM system was calibrated by using a 1 mm/100 divisions stage micrometer. The actual magnification of the system is given as 42×. The lateral resolution of the system was found to be 0.11 μm and the minimum axial resolution of the system is 40 nm. The measurement speed of the system is determined by the capture speed of the CCD camera at 15 frames/s.

A refractive quartz microlens array from SUSS MicroOptics with a 100-μm pitch was tested by the DHM system described above. As discussed above, if there is a system phase when numerical compensation is used, there may be false compensation of the numerical phase of the spherical lens. Here, we present some examples as illustration. The microlens was measured with different reference waves. Figure 3.35a shows the system phase in the case of misalignment of the DHM system. When the test microlens was introduced into the system, the phases are as shown in Figure 3.35b. In such cases, numerical phase compensation can easily give rise to false measurement results when only a single hologram is employed.

We modified the reference wave to achieve a quasi-flat phase. When the same test microlens was introduced into the system, the intensity obtained is shown in Figure 3.36a. The wrapped phase is shown in Figure 3.36b. Given that the refractive index is 1.457 at 633 nm, the 3D height display of the lens is shown in Figure 3.36c.

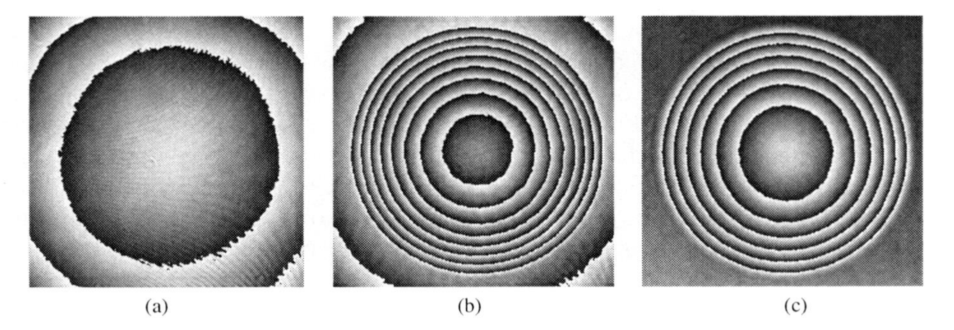

Figure 3.35 Microlens with different system phase. (a) System phase resulting from difference between reference waves and object waves; (b) Phase of the microlens in addition to the system phase shown in (a); (c) Phase of the microlens for comparison

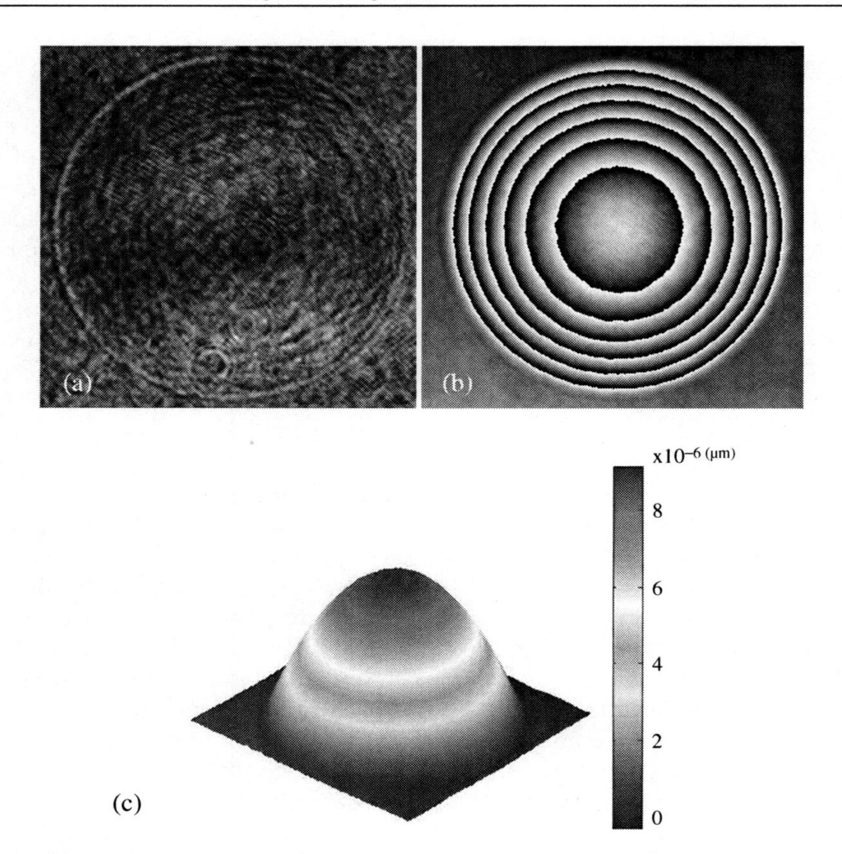

Figure 3.36 Microlens characterization. (a) Intensity; (b) Wrapped phase; (c) 3D display of the unwrapped phase of (b)

A refractive quartz microlens array from SUSS MicroOptics with a 22-μm pitch was tested by the same DHM system. The spectrum distribution of the hologram without the test object is shown in Figure 3.37a. When the test object was introduced, the spectrum of interest showed a dispersive distribution as shown in Figure 3.37b. In such a case, it is difficult to locate the center of the spectrum. Figure 3.38a shows an incompletely removed off-axis tilt resulting from placing the spectrum of interest off-center. Manually removing the off-axis tilt is feasible but very time-consuming.

To solve the problem, there is a robust method to perform the reconstruction of the hologram. One can fix the center of the spectrum of interest based on the DHM system without the test specimen in the numerical reconstruction process. Since the center is fixed and the system phase is removed, there will be no off-axis tilt and spherical phase in the

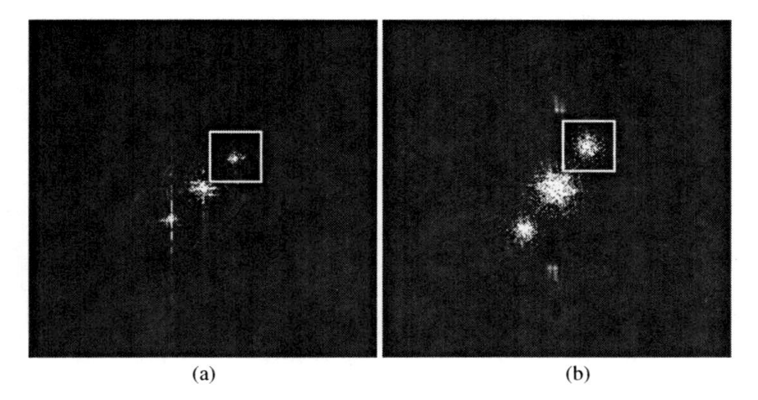

Figure 3.37 (a) The frequency spectra of a digital hologram with a flat phase surface; (b) The frequency spectra of a digital hologram of a microlens array

final phase measurement. One can then insert the sample to be tested and measure only the phase introduced by the sample itself. As shown in Figure 3.38b, the off-axis tilt has been removed to give the correct phase measurement. Given the refractive index of 1.457 at 633 nm, the 3D height map of the lens is shown in Figure 3.39.

The living Vero cells were used as the test specimen. 100% confluent Vero cells from a T25 flask were trypsinized and diluted ten times into two T25 flasks to a final culture volume of 10 ml. The flasks were incubated at 37 °C with 5% CO_2 for 12 hours to allow adherence of monolayer cells. Images were captured at 60-second intervals through a 12-hour period at room temperature.

A quantitative phase map of Vero cell division is achieved as shown in Figure 3.40. At t = 27 minute, the cell of interest is in metaphase as shown in Figure 3.41a. At t = 37 minute, it starts its anaphase as shown in Figure 3.41b.

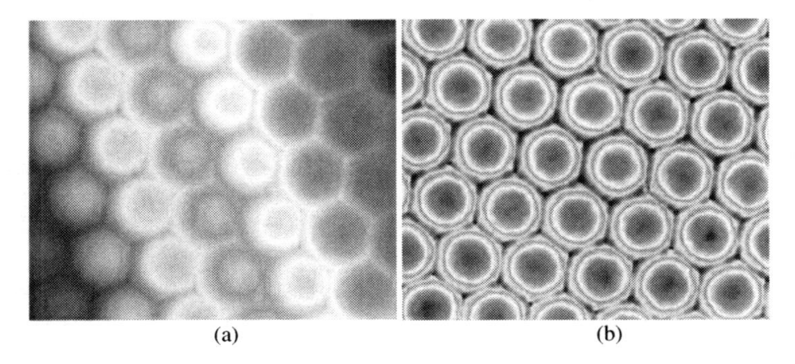

Figure 3.38 Phase of the microlens array. (a) With off-axis tilt incompletely removed; (b) Without off-axis tilt

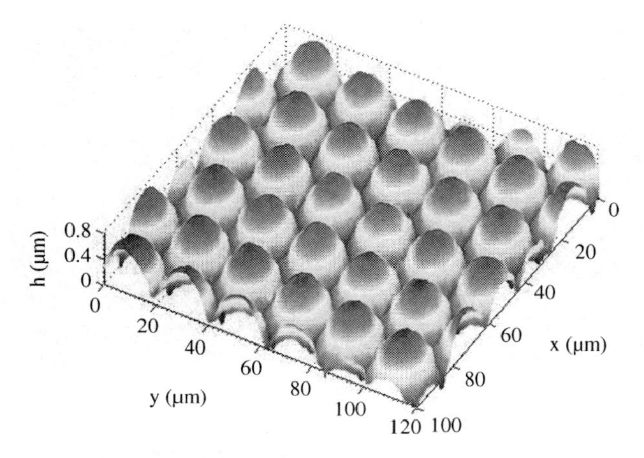

Figure 3.39 3D height map of the microlens array

Figure 3.40 Quantitative phase map of living Vero cell during the cell division (a) Metaphase; (b) and (c) Anaphase; (d) and (e) Cytokinesis; (f) Separation of the two sister cells

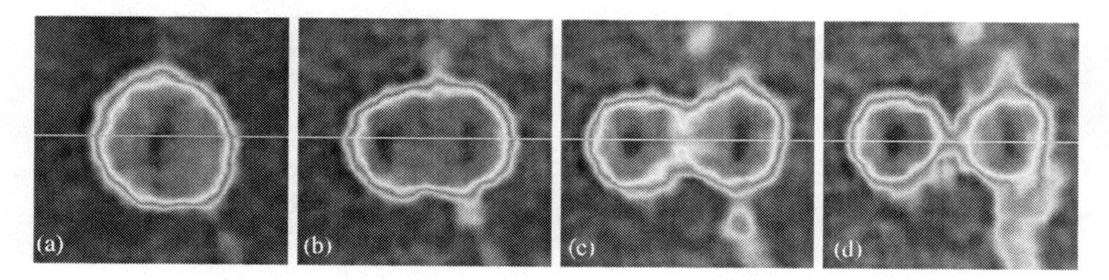

Figure 3.41 Monitoring of living Vero cell during the cell division (a) t = 27 minute; (b) t = 37 minute; (c) t = 41 minute; (d) t = 43 minute

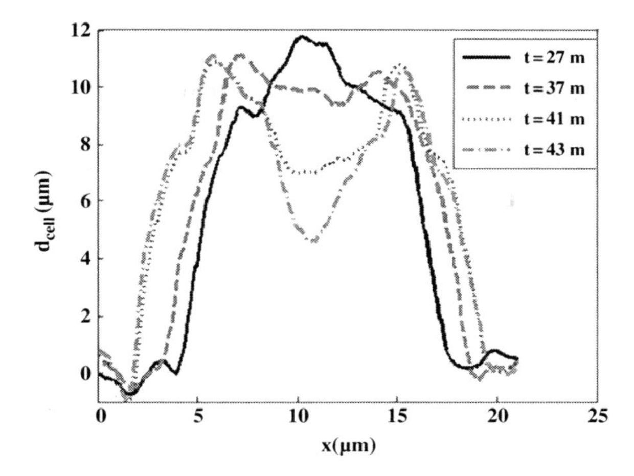

Figure 3.42 Cross-sections through the cell thickness given by the phase marked by the solid line in Figure 3.4

At t = 41 minute, it starts its cytokinesis as shown in Figure 3.41c. At t = 43 minute, the separation of the two sister cells is shown in Figure 3.41d. Figure 3.42 shows cross-sections of the cell thickness obtained from the measured optical path-length changes with the refractive index of the medium 1.33, the Vero cell 1.39 under the wavelength 633 nm.

3.3.5 Digital Holographic Microscopy with Quasi-Physical Spherical Phase Compensation : Light with Short Coherence Length (47)

The other proposed DHM systems is also built in the Michelson interferometer configuration which uses an adjustable lens in the reference beam to perform the quasi-physical spherical phase compensation. It is built to use a light source with a short coherence length. As shown in Figure 3.43, an adjustable lens is used in the reference beam path. The coherence length of the light source may limit the position adjustable capability of the reference wave. In such a case, the adjustable lens can be used to change the phase curvature to fulfill the spherical phase compensation in the hologram recording process.

Because of the short coherence length, the distance between the object source and the hologram recording plane must be the same as the distance between the reference source and the hologram recording plane. The light beam for the object path goes through the condenser, the microscope objective and the tube lens. Consequently the wavefront has changed. There is a difference between the object wavefront and the reference wavefront at the hologram recording plane and this is shown by the hologram in Figure 3.44a. The space frequency of the fringe is very high. We can analyze the Fourier spectra and find the

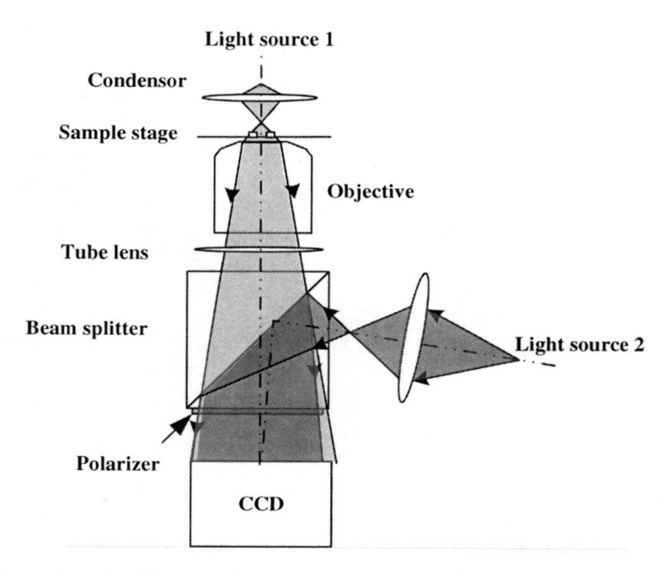

Figure 3.43 Schematic of transmission mode DHM set-up for short coherence light source

difference between the object wavefront and the reference wavefront. The Fourier spectra are shown in Figure 3.44b. The spectrum of the first order is rectangular due to the shape of the CCD sensor. We can find the center of the spectrum and move it to the center of the calculation coordinate in order to remove the off-axis tilt. Consequently we can extract the phase from the digital hologram as shown in Figure 3.44c. It is a spherical phase which is called a system phase in the previous sections. Actually it is the difference between the object wavefront and the reference wavefront at the hologram recording plane. This spherical phase can be removed by introducing a lens in the reference path. The lens is used to change the reference wavefront. When the position of the lens is changed, the reference wavefront is also changed. The real-time phase monitoring software can tell the exact position of the lens to make sure of the physical spherical phase compensation. There is one and only one position for a certain lens to fulfill the physical spherical phase compensation. The hologram is recorded as shown in Figure 3.44d. The Fourier spectra are shown in Figure 3.44e. The spectrum of the first order becomes a point to indicate there is no difference between the object wavefront and the reference wavefront. The flat phase extracted from the hologram is shown in Figure 3.44f.

It should be mentioned that a plane numerical reference wavefront without any additional phase introduction to the reconstructed object phase is used in the reconstruction process. Consequently, under the physical spherical phase compensated recording condition, the phase extracted from the recorded

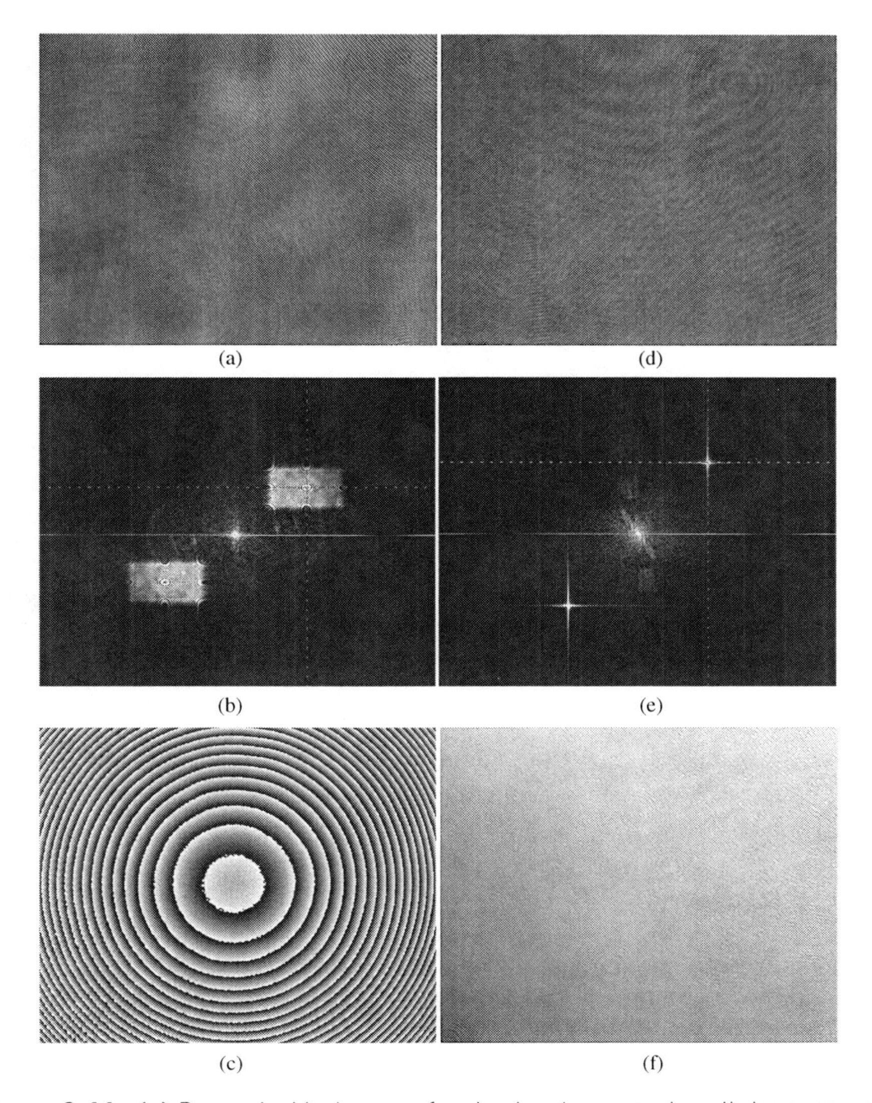

Figure 3.44 (a) Recorded hologram for short coherence length laser source; (b) Fourier spectra of the hologram (a); (c) Reconstructed phase from the selected order marked with dotted cross line in (b); (d) Recorded hologram when use lens for physical spherical phase compensation; (e) Fourier spectra of the hologram (d); (f) Reconstructed phase from the selected order marked with dotted cross line in (e)

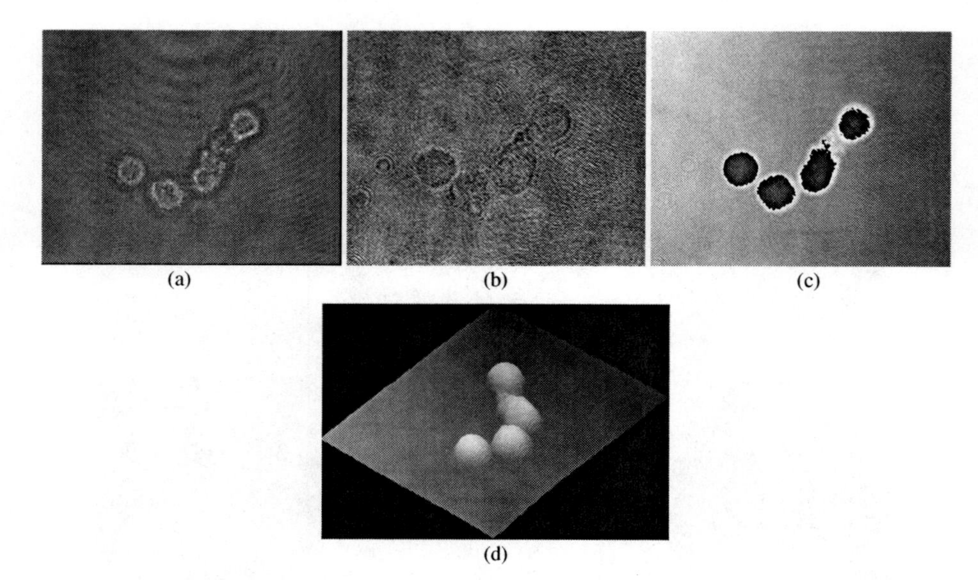

Figure 3.45 Application to cell morphology. (a) Hologram; (b) Reconstructed intensity from (a); (c) Reconstructed phase from (a); (d) The unwrapped 3D phase map of (c)

hologram is only introduced by the test sample. To demonstrate, the cultured living yeast cell is measured by the microscope system. The results are shown in Figure 3.45.

3.4 Conclusion

Digital holography is based on diffraction and interference of the light wave. It has the same foundation as classic holography in the hologram recording process. Numerical reconstruction makes digital holography capable of phase acquisition and display. Thus, both an optical set-up and computer software are needed in digital holography.

In order to solve the problem of resolution limitation by the current digital capture device, DHM became increasingly popular in quantitative phase microscopy. The imaging resolution is greatly improved by using a microscope objective. At the same time the phase of the light wave is changed by the introduction of the microscope objective. This makes the numerical reconstruction more complicated than ever before. When a plane wave is taken as the reference recording wave, the original wavefront can always be recorded as a spherical wavefront. When a spherical wave is taken as the reference recording wave, there still is a system spherical phase curvature as with different recording conditions. A detailed interference analysis between two diverging wavefronts shows that there is an optimized recording condition for a digital hologram.

When the two involved diverging point sources are located at the positions where there is a relatively equal distance from the hologram recording plane, the spherical phase of each wavefront will be compensated by each other. A flat phase can be extracted from the recorded hologram by using a numerical plane reference waves. This is different from the classical holography in the reconstruction reference waves. The former uses numerical plane reference waves, the latter uses the original recording reference wave or its conjugation. It is easy to understand that the original recording reference wave will reconstruct the original object wave. But for DHM, the use of microscope objective makes the object wave a spherical phase curvature. This spherical curvature gives a distortion measurement phase result. Numerical phase compensation actually is made to compensate such a spherical phase. It has been proved successful for many applications. But if the phase introduced by the test object itself has a spherical phase curvature, numerical phase compensation may give a false measurement result by under-compensation or over-compensation for the spherical phase.

Physical spherical phase compensation for classical holography is expensive because the same optical component is needed in the two optical paths of the interferometer. It is also difficult to precisely match the two optical paths without any visual direction. The common-path interferometer solves the two problems at the same time. The use of the single cube beam splitter to separate and recombine the two optical paths to give an off-axis digital hologram not only simplifies the optical system but also easily achieves the physical spherical phase compensation. Transmission DHM is based on the lens-less concept and a lens concept is proposed using such a common-path interferometer. The experimental demonstration shows that they can be usefully applied to the quantitative phase measurement of microscopic object. The common-path DHM is very suitable for the linear microlens array characterization and the fast uniformity inspection of the whole array. Microlens characterization and microlens array characterization are used to demonstrate the measurement capability of the DHM.

Ways of quasi-spherical phase compensation are proposed based on the Michelson interferometer. The merit of it is the easy adjustment of the two optical path lengths. A real-time phase monitoring software is necessary to direct the adjustment of the relative position of the light source. For a light source with a long coherence length (an He-Ne laser), the optical path length can be different to give a good interference pattern. Consequently one can fix the imaging optical path then adjust the reference beam path to quasi-match the curvature of the optical wavefront. If a light source directly emitted from the fiber points is used, it is not necessary to insert an additional lens to change the wavefront. A simple method is to adjust the position of the fiber point and change the wavefront curvature reaching the hologram recording plane. There is only one position for the reference fiber point to achieve a quasi-spherical phase compensated digital hologram.

For a light source with a short coherence length (a laser diode), the optical path length of the two optical beams must be equal to achieve the best interference fringe pattern. Since the object beam path and the reference beam path pass through different optical components, it is hard to obtain the same curvature in the hologram recording plane without an additional lens in the reference beam path. Consequently a lens or a lens combination can be used in the reference beam path to change the wavefront curvature and give a quasi-match for the object wavefront. Thus quasi-physical spherical phase compensation can be achieved.

Physical spherical phase compensation in transmission DHM makes the quantitative phase measurement by only a single hologram faster and easier than that of numerical phase compensation. It provides a way of simplifying not only the optical system but also the numerical reconstruction. The lateral resolution of the DHM can be improved by using the microscope objective with a high magnification. The phase measurement accuracy depends on the beam irradiance distribution.

References

1. Gabor, D. (1949) Microscopy by reconstructed wave-fronts. *Proc. Roy. Soc. Lond. A Mat.*, **197**, 454–487, http://www.jstor.org/stable/98251.
2. Silverman, B.A., Thompson, B.J., and Ward, J.H. (1964) A laser fog disdrometer. *J. Appl. Meteorol.*, **3**, 792–801.
3. Thompson, B.J., Parrent, G.B., Ward, J.H., and Justh, B. (1966) A readout technique for the laser fog disdrometer. *J. Appl. Meteorol.*, **5**, 343–348.
4. Hobson, P.R. (1988) Precision coordinate measurements using holographic recording. *J. Phys. E: Sci. Instrum.*, **21**, 139–145.
5. Trolinger, J.D., Belz, R.A., and Farmer, W.M. (1969) Holographic techniques for the study of dynamic particle fields. *Appl. Opt.*, **8**, 957–961.
6. Meng, H., Anderson, W.L., Hussain, F., and Liu, D.D. (1993) Intrinsic speckle noise in in-line particle holography. *J. Opt. Soc. Am. A.*, **10**, 2046–2058.
7. Thompson, B.J., Ward, J.H., and Zinky, W.R. (1967) Application of hologram techniques for particle size analysis. *Appl. Opt.*, **6**, 519–526.
8. Gabor, D. and Goss, W.P. (1966) Interference microscope with total wavefront reconstruction. *J. Opt. Soc. Am. A.*, **56**, 849–856.
9. Schnars, U. and Jüptner, W. (2002) Digital recording and numerical reconstruction of holograms. *Meas. Sci. Technol.*, **13**, R85–R101.
10. Leith, E.N. and Upatnieks, J. (1962) Reconstructed wavefronts and communication theory. *J. Opt. Soc. Am. A.*, **52**, 1123.

11. Bryngdah, O. and Lohmann, A. (1968) A. single-sideband holography. *J. Opt. Soc. Am.*, **58**, 620.
12. Kronrod, M.A., Yaroslavski, L.P., and Merzlyakov, N.S. (1972) Computer synthesis of transparency holograms. *Sov. Phys. Tech. Phys.*, **17**, 329–332.
13. Onural, L. and Scott, P.D. (1987) Digital decoding of in-line holograms. *Opt. Eng.*, **28**, 1124.
14. Liu, G. and Scott, P.D. (1987) Phase retrieval and twin-image elimination for in-line Fresnel holograms. *J. Opt. Soc. Am. A.*, **4**, 159–165.
15. Jüptner, U.S.a.W. (1994) Direct recording of holograms by a CCD-target and numerical reconstruction. *Appl. Opt.*, **33**, 179–181.
16. Yamaguchi, I. and Zhang, T. (1997) Phase-shifting digital holography. *Opt. Lett.*, **22**, 1268–1270.
17. Zhang, T. and Yamaguchi, I. (1998) Three-dimensional microscopy with phase-shifting digital holography. *Opt. Lett.*, **23**, 1221–1223.
18. Javidi, B. and Tajahuerce, E. (2000) Three-dimensional object recognition by use of digital holography. *Opt. Lett.*, **25**, 610–612.
19. Yamaguchi, I., Ohta, S., and Kato, J. (2001) Surface shape measurement by phase-shifting digital holography. *Opt. Rev.*, **8**, 85–89.
20. Kostianovski, S., Lipson, S.G., and Ribak, E.N. (1993) Interference microscopy and Fourier fringe analysis applied to measuring the spatial refractive-index distribution. *Applied Optics*, **32**, 7.
21. Haddad, W., Cullen, D., Solem, J.C. *et al.* (1992) Fourier-transform holographic microscope. *Appl. Opt.*, **31**, 4973–4978.
22. Cuche, E., Bevilacqua, F., and Depeursinge, C. (1999) Digital holography for quantitative phase-contrast imaging. *Opt. Lett.*, **24**, 291.
23. Zhang, F., Yamaguchi, I., and Yaroslavsky, L.P. (2004) Algorithm for reconstruction of digital holograms with adjustable magnification. *Opt. Lett.*, **29**, 1668–1670.
24. Nicola, S.D., Finizio, A., Pierattini, G. *et al.* (2005) Angular spectrum method with correction of anamorphism for numerical reconstruction of digital holograms on tilted planes. *Opt. Express*, **13**, 9935–9940, http://www.opticsinfobase.org/oe/abstract.cfm?uri=oe-13-24-9935.
25. Yu, L. and Kim, M.K. (2006) Pixel resolution control in numerical reconstruction of digital holography. *Opt. Lett.*, **31**, 897–899.
26. Ferraro, P., Nicola, S.D., Coppola, G. *et al.* (2004) Controlling image size as a function of distance and wavelength in Fresnel-transform reconstruction of digital holograms. *Opt. Lett.*, **29**, 854–856.
27. Malacara, D. (ed.) (1992) *Optical Shop Testing*, John Wiley & Sons.
28. Mann, C., Yu, L., Lo, C.-M., and Kim, M. (2005) High-resolution quantitative phase-contrast microscopy by digital holography. *Opt. Express*, **13**, 8693–8698.
29. Ya'nan, Z., Weijuan, Q., De'an, L. *et al.* (2006) Ridge-shape phase distribution adjacent to 180° domain wall in congruent LiNbO3 crystal. *Appl. Phy. Leet.*, **89**, 112912.
30. Gaskill, J.D. (ed.) (1978) *Linear Systems, Fourier Transforms, and Optics*, John Wiley & Sons.
31. Cuche, E., Marquet, P., and Depeursinge, C. (2000) Spatial filtering for zero-order and twin-image elimination in digital off-axis holography. *Appl. Opt.*, **39**, 4070.

32. Montfort, F., Charrière, F., Colomb, T. *et al.* (2006) Purely numerical compensation for microscope objective phase curvature in digital holographic microscopy: influence of digital phase mask position. *J. Opt. Soc. Am. A*, **23**, 2944.
33. Schnars, U. (1994) Direct phase determination in hologram interferometry with the use of digitally recorded holograms. *J. Opt. Soc. Am. A*, **11**, 2011–2015.
34. Marquet, P., Cuche, E., Depeursinge, C. *et al.* (2005) Apparatus and methods for digital holographic imaging, US Patent, NO. US 6943924B2.
35. Cuche, E., Marquet, P., and Depeursinge, a.C. (1999) Simultaneous amplitude-contrast and quantitative phase-contrast microscopy by numerical reconstruction of Fresnel off-axis holograms. *Appl. Opt.*, **38**, 6994–7001.
36. Colomb, T., Montfort, F., Kühn, J. *et al.* (2006) Numerical parametric lens for shifting, magnification, and complete aberration compensation in digital holographic microscopy. *J. Opt. Soc. Am. A*, **23**, 3177.
37. Colomb, T., Cuche, E., Charrière, F. *et al.* (2006) Automatic procedure for aberration compensation in digital holographic microscopy and applications to specimen shape compensation. *Appl. Opt.*, **45**, 851.
38. Miccio, L., Alfieri, D., Grilli, S., Ferraroa, P., Finizio, A., Petrocellis, L. D., and Nicola, S.D. (2007) Direct full compensation of the aberrations in quantitative phase microscopy of thin objects by a single digital hologram. *Appl. Phy. Lett.*, **90**, 041104–041101–041103.
39. Stadelmaier, A. and Massig, J.H. (2000) Compensation of lens aberrations in digital holography. *Opt. Lett.*, **25**, 3.
40. Carl, D., Kemper, B., Wernicke, G., and Bally, G.v. (2004) Parameter-optimized digital holographic microscope for high-resolution living-cell analysis. *Appl. Opt.*, **43**, 9.
41. Ferraro, P., Nicola, S.D., Finizio, A. *et al.* (2003) Compensation of the inherent wave front curvature in digital holographic coherent microscopy for quantitative phase-contrast imaging. *Appl. Opt.*, **42**, 1938–1946.
42. Di. J., Zhao, J., Sun, W., Jiang, H., and Yan, X. (2009) Phase aberration compensation of digital holographic microscopy based on least squares surface fitting. *Opt Commun.*, **282**(19): 3873–3877.
43. Kühn, J., Charrière, F., Colomb, T. *et al.* (2008) Axial sub-nanometer accuracy in digital holographic microscopy. *Meas. Sci. Technol.*, **19**, 074007.
44. Pavillon, N., Seelamantula, C. S., Kühn, J. *et al.* (2009) Suppression of the zero-order term in off-axis digital holography through nonlinear filtering *Appl. Opt.*, **48**, H186–H195.
45. Colomb, T., Kühn, J., Charrière, F. *et al.* (2006) Total aberrations compensation in digital holographic microscopy with a reference conjugated hologram. *Opt. Express*, **14**, 4300.
46. Mann, C.J., Yu, L., Lo, C.-M., and Kim, A.M.K. (2006) High-resolution quantitative phase-contrast microscopy by digital holography. *Opt. Express*, **13**, 8693–8698.
47. Weijuan, Q., Choo, C.O., Singh, V.R. *et al.* (2009) Quasi-physical phase compensation in digital holographic microscopy. *J. Opt. Soc. Am. A*, **26**, 2005–2011.
48. Kemper, B. and Bally, G.v. (2008) Digital holographic microscopy for live cell applications and technical inspection. *Appl. Opt.*, **47**, 10.

49. Qu, W., Bhattacharya, K., Choo, C.O. *et al.* (2009) Transmission digital holographic microscopy based on a beam-splitter cube interferometer. *Appl. Opt.*, **48**, 2778–2783.
50. Ferrari, J.A. and Frins, E.M. (2007) Single-element interferometer. *Opt. Commun.*, **279**, 235–239.
51. Qu, W., Yu, Y., Chee, O.C., and Asundi, A. (2009) Digital holographic microscopy with physical phase compensation. *Opt. Lett.*, **34**, 1276–1886–890278.
52. Qu, W., Choo, C.O., Yingjie, Y., and Asundi, A. (2011) Characterization and inspection of microlens array by a single cube beam splitter microscopy. *Appl. Opt.*, **50**, 886–890.
53. Qu, W., Choo, C.O., Yingjie, Y., and Asundi, A. (2010) Microlens characterization by digital holographic microscopy with physical spherical phase compensation. *Appl. Opt.*, **49**, 6448–6454.

4

Digital In-Line Holography and Applications

Taslima Khanam

4.1 Background

Particle and powder technology appears one of the major contribution in modern science and engineering. Industries such as chemical, pharmaceutical, food, mineral, hydro-fuel and so on deal with particulate matter, both in dry and wet form, for example, powder, paste, spray, emulsion, and others. Quality production of these industries often relies on the controlled size and shape of the particles. Hence, it is very important to accurately measure the particle size and shape to determine and control the quality of the products. Hence, the development of an appropriate tool for particle size and shape measurement is increasingly of interest to researchers.

In practical situations, particles are contained in volume, and the size of the particles is of the order of micrometer resolution. These are the challenges commonly encountered in developing the tool for particle sizing. Most of the tools that analyze particles are not completely satisfactory in meeting both challenges. Conventional microscopy or imaging tool often meets the micrometer resolution but becomes problematic for the study of the volume of the particle field, as these tools suffer from the reduced depth of field imposed by the required magnification. As a result, they can only be used to study particles only on a plane. Thus conventional imaging tools are inadequate to determine

Digital Holography for MEMS and Microsystem Metrology, First Edition. Edited by Anand Asundi.

the perspective position of particles in volume, the study of their random orientation, and the dynamic volume of a particle field [1]. Apart from the imaging tool, there are several non-imaging tools, such as light scattering, diffraction methods [2]. These tools are good enough for rapid and mean particle size distribution where exact shape and other parameters are not needed. But these techniques depend on models which require the assumptions of particle shape and pre-knowledge of other physical parameters, and most of these techniques assume that the particles are perfect spheres. Hence, model-based measurement methods are valuable in particular situations, but there is often a mismatch between ideal and practical situations due to the presence of particles with irregular or unknown geometry. As far as both size and shape are concerned, obviously some kind of imaging tool is needed that is not limited to just a plane, but rather is capable of storing the whole volume information with good resolution capability.

Holography is a novel tool that helps overcome the limitations of classical imaging techniques. It uses a coherent light source and records the interference between the un-diffracted (reference wave) and the diffracted (object wave) light from an object. This interference pattern contains information about the entire volume of the sample compared to conventional imaging techniques that record information only from the focused plane. Holographic particle analysis is an established technique for particle measurement [3]. The use of high definition photographic films for the recording of particle holograms allows the study of samples with relatively large volumes and with high resolution but suffers from increased processing time required for film development and subsequent digitization of the reconstructions. However, with the digital recording of holograms the chemical processing step for the film development is eliminated which substantially increases the practicability of the method for on-line application. The resolution achievable by digital holographic systems might not be as high as the one obtained with holograms recorded on high resolution films, but the ease of processing has made digital holography more attractive for particle analysis. Among the various set-ups for geometry of digital holography, digital in-line holography is mostly used for the application of particle analysis [4–8]. In comparison to the off-axis set-up, in-line geometry helps to relax the spatial resolution requirement on the CCD sensor, provides higher imaging resolution and larger field of view, and thereby is extremely useful in the study of particle field [9].

The aim of this chapter is to study the application of lens-less digital in-line holography for particle measurement. Detailed methodologies for measuring particle size and shape are discussed here. Particles studied in this chapter are, spheres and needle-like ones that represent the extreme shape characteristics of particles encountered in practice. Subsequent validation and performance of the digital in-line holography (DIH) based method for dry, wet and flowing particles are also studied in this chapter.

4.2 Digital In-Line Holography

Holography is an optical phenomenon that originated from the fundamental theory of diffraction and interference. This method is capable of recording and reconstructing the optical wave fields [3, 10]. When a coherent light illuminates an object, it diffracts light. The interference between the diffracted light (object wave) and the source light (reference wave) constitutes the hologram. Unlike the most commonly used off-axis approach, the in-line approach of holography uses one beam as in Figure 4.1a. In the arrangement shown in Figure 4.1a, only a single beam is used to illuminate the object, hence the diffraction by the object cross-section acts as an object beam and the directly transmitted light acts as a reference beam. Thus the object and reference wave propagate along the same optical axis, and constitute the in-line hologram. When the hologram is reconstructed, only the object cross-section is reconstructed, and twin images of the object and directly transmitted light will appear along the same line as shown in Figure 4.1b. In the following section, details of the recording and reconstruction procedure of a digital in-line hologram are discussed.

4.2.1 Recording and Reconstruction

In digital holography, the hologram constituting the interference between the plane reference wave and the waves scattered by the object is directly recorded

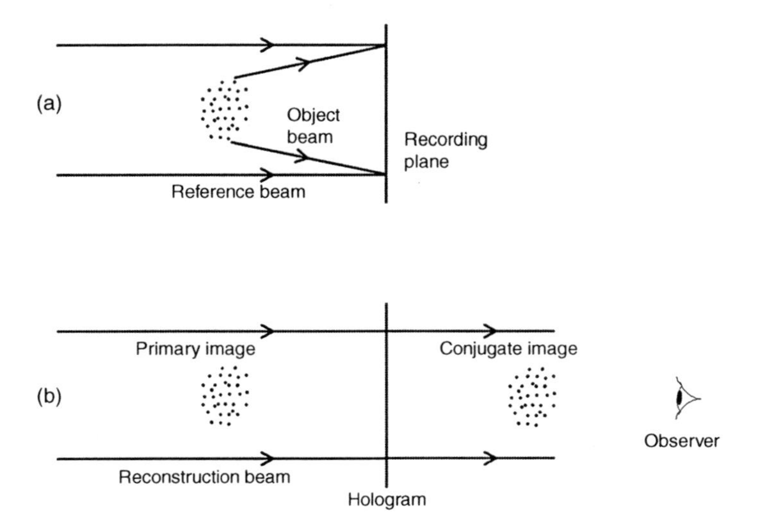

Figure 4.1 In-line holography with a single beam. (a) Recording with a transmitted or diffracted object field; (b) Reconstruction

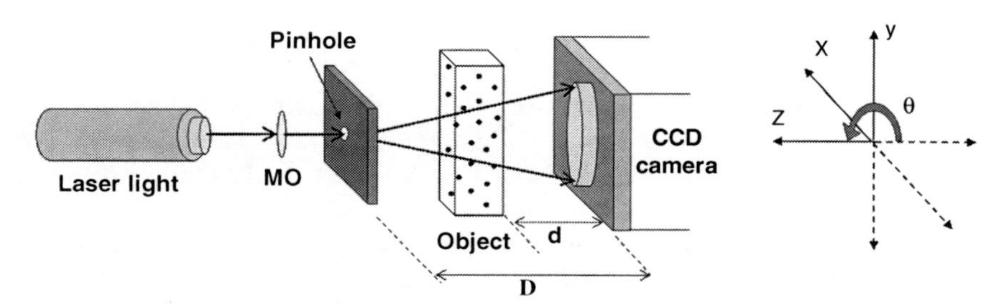

Figure 4.2 Lens-less digital in-line holography set-up

and stored as a digital image at the surface of the charge coupled device (CCD) without any focusing optics. The set-up for the recording of the digital holograms is shown in Figure 4.2. A laser beam is focused by a microscopic objective lens onto a pinhole, which is located at a distance D from the recording camera. The resulting spherical diverging beam originating from the pinhole illuminates the object, a diffusely reflecting body which is located at a distance "*d*" from the recording camera. One part of the beam passes through the object without being diffracted and acts as the reference beam U_R. Another part of the beam is diffracted by the object, generating the object beam U_d, which propagates towards the recording camera. The resulting interference between the reference and the object waves constitutes the hologram on the CCD plane. The hologram can be used to reconstruct the recorded scene at the desired distance, d, from the recording plane using the Fresnel-Kirchhoff integral [10]. Thus, from a single recorded hologram, objects at different positions along the optical axis (Z-axis) within the recording depth, D, can be numerically focused by changing d. It is noted that the diverging beam introduces the magnification in the set-up which is a function of the reconstruction distance d. Hence, the set-up shown in Figure 4.2 is referred as lens-less digital in-line holographic microscopy (LDIHM). The advantage of the lens-less system is that it is free of lens aberration.

The spherical diverging reference beam $\mathrm{U_R}$ can be expressed as shown in [10] as:

$$U_R(x,y) = \frac{\exp\left(-i\frac{2\pi}{\lambda}\sqrt{x^2+y^2+D^2}\right)}{\sqrt{x^2+y^2+D^2}} \tag{4.1}$$

where x and y are the spatial coordinates on the camera plane, and λ is the wavelength of the laser beam and $i=\sqrt{-1}$. The term $\sqrt{(x^2+y^2+D^2)}$

describes the distance between the point source and any point on the camera plane. The object beam U_d at the camera plane can be expressed by the Fresnel-Kirchhoff integral as shown in [11],

$$U(x,y) = \int_{-\infty}^{+\infty}\int_{-\infty}^{+\infty} U_d(x',y') \frac{\exp\left[-i\frac{2\pi}{\lambda}\sqrt{(x'-x)^2+(y'-y)^2+d^2}\right]}{\sqrt{(x'-x)^2+(y'-y)^2+d^2}} dx'dy', \quad (4.2)$$

where x', y' are the spatial coordinates on the object plane. The term $\sqrt{(x'-x)^2+(y'-y)^2+d^2}$ describes the distance between a point on the object plane and a point on the camera plane. For simplicity, constant phase terms have been ignored in Equation 4.2. The resulting hologram captured by a CCD camera can be expressed as:

$$I(x,y) = [U_R(x,y) + U(x,y)]^2 \quad (4.3)$$

For the reconstruction of the hologram at a distance d', I has to be multiplied by the reference wave and then propagated using the Fresnel-Kirchhoff integral as given in [10]

$$U_d(x',y') = \int_{-\infty}^{+\infty}\int_{-\infty}^{+\infty} I(x,y)U_R(x,y) \frac{\exp\left[-i\frac{2\pi}{\lambda}\sqrt{(x-x')^2+(y-y')^2+d'^2}\right]}{\sqrt{(x-x')^2+(y-y')^2+d'^2}} dxdy \quad (4.4)$$

For $d'=d$ and when quantization and other digitizing errors are negligible, the reconstructed wave $U_{d'}$ equals the object wave U_d.

In digital holographic microscopy, Equation 4.4 has to be numerically calculated with the convolution reconstruction method, in which case, the pixel size of the reconstructed image ($\Delta x'$, $\Delta y'$) equals the pixel size of the recording camera (Δx, Δy), that is, $\Delta x' = \Delta x$ and $\Delta y = \Delta y'$. Lateral magnification defined as $M(d) = r'(d)/r$, where $r'(d)$ is the measured object size at distance d and r is the real object size, can be introduced by changing the distance between the source of the spherical reference wave and the CCD, or the wavelength that is used for the reconstruction as given in [10]

$$M(d) = \left(1 + \frac{d}{D'}\frac{\lambda}{\lambda'} - \frac{d}{D}\right)^{-1} \quad (4.5)$$

where D' is the pinhole to CCD distance and λ' is the wavelength that are used for the reconstruction, respectively. By changing D' and λ', the distance d' where the object will appear focused also changes as given in [10]

$$d' = \left(\frac{1}{D'} + \frac{\lambda'}{\lambda} \frac{1}{d} - \frac{1}{D} \frac{\lambda'}{\lambda} \right)^{-1} \tag{4.6}$$

In this chapter, all the experiments use $D' = 100D$, and $\lambda' = \lambda$.

The digital holography set-up which has been described above has three desirable characteristics. First, it can magnify the reconstructed object wave, allowing the study of smaller particles compared to systems which use collimated illumination. Second, it minimizes the effects of the twin image since the twin image is well out of focus for the depths where the particles appear focused [11]. Finally, the described reconstruction method simplifies the spatial localization of the particles as it overcomes the pixel resizing problem which causes a shift of the *x*-*y* location of the particles for different depth reconstructions.

4.3 Methodology for 2D Measurement of Micro-Particles

In this section, an overview of how to process the hologram and algorithm to extract the particle size, shape and focusing depth from the recorded holograms is provided [4–6]. A summary of the processing algorithm is illustrated in Figure 4.3. Brief details are discussed as follows.

4.3.1 Numerical Reconstruction, Pre-Processing and Background Correction

In order to study a volume of the sample, first, each hologram needs to be reconstructed at several reconstruction depths. Reconstructions are carried out using the method in Section 4.2.1 covering the volume to be studied with sufficiently small depth steps, because the distance between successive reconstructions depends on the minimum size of the particles to be studied. Hence, several particles will appear in each reconstructed image as either focused or out of focused particles. A sample hologram and its reconstructed image are shown in Figures 4.4a and b, respectively. As shown in Figure 4.4b, the background of the reconstructed hologram is not uniform. The bright background is due to the zero order of the reconstruction which is also contaminated by noise caused by speckle due to the coherent illumination, the twin image and non-uniform illumination. One approach to reduce the zero-order term from the reconstruction is to high-pass the hologram with a low cut-off frequency filter before reconstruction [10]. Another approach is to record the reference

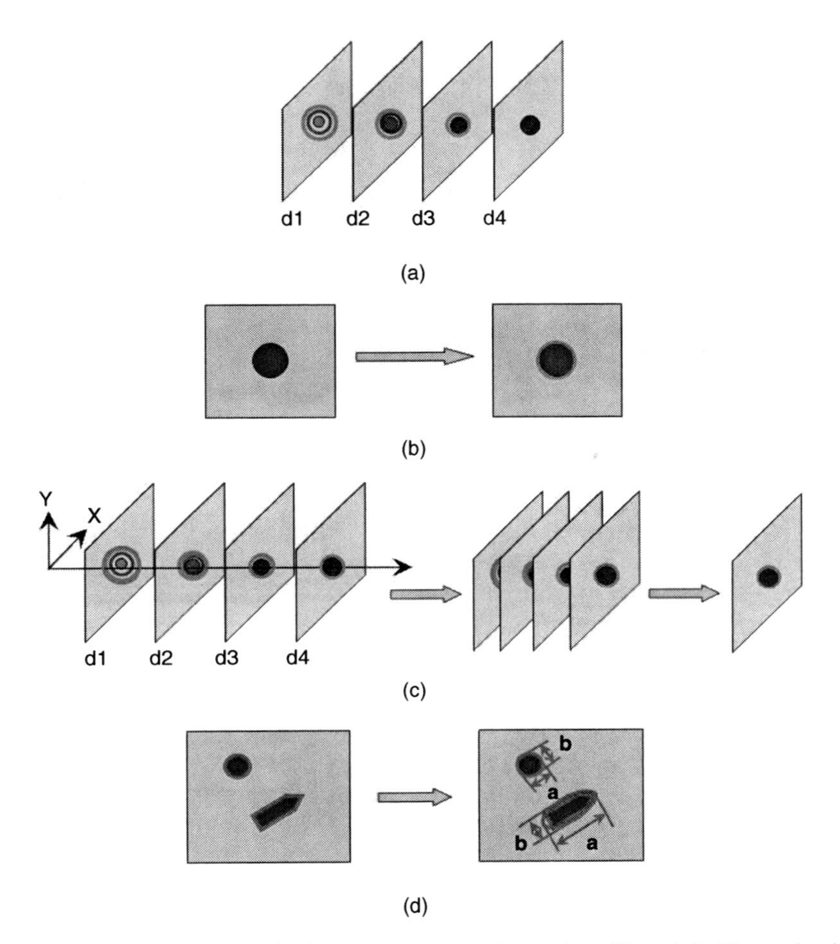

Figure 4.3 Steps in the hologram processing algorithm.(a) Numerical reconstruction at several depths; (b) Image segmentation using Canny edge detection; (c) Localization of a particle; (d) Particle size and shape measurement *Note*: d_{eq} denotes the equivalent diameter for spherical objects, while m_1 and m_2 represent the major and minor axes for non-spherical particles (Reproduced by permission of © 2009 Elsevier)

wave (without the object) and subtract it from the hologram prior to reconstruction [7]. This approach, however, is impractical for real-time monitoring systems as the sample needs to be removed from the system regularly in order to record updated images of the reference wave. In all cases, we have noticed that the suppression of the zero-order term affects the particles, by blurring or changing their perimeter, leading to erroneous size measurements. As a result, these approaches offer practical disadvantages. In the following

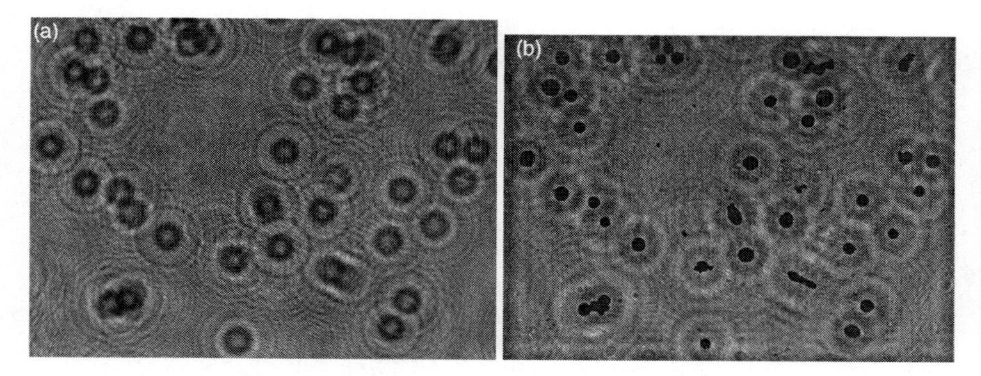

Figure 4.4 (a) Hologram of ceramic beads on glass slide; (b) Example of a reconstructed image (Reproduced by permission of © 2009 Elsevier)

section, we propose an edge detection-based technique for particle segmentation which is immune to zero order and other irregularities of the background, thus eliminating pre-processing of the hologram.

4.3.2 Image Segmentation

After numerical reconstructions, Canny edge detection is applied on each reconstructed image to identify particles from the images [12]. This edge detection technique locates edges by examining gradients after a Gaussian filter has been applied to the reconstruction. Gaussian filtering reduces the effect of noise on erroneous identification of edges. Following Gaussian filtering, the edge detection algorithm examines the derivatives for local maxima. Thresholding with hysteresis is used to classify the identified maxima. Two thresholds t_L and t_H are used for this. Maxima with a value lower than t_L are identified as not being edges immediately. Maxima with a value higher than t_H are identified as edges immediately. Maxima with value between t_L and t_H are identified as edges only if they are connected to identified edges. The algorithm results in a set of ones ("1") where edges have been detected and zeros everywhere else.

The method requires the selection of three parameters: The standard deviation σ and the two thresholds t_L and t_H. Standard deviation, σ is selected so that the filtering operation does not alter the size of the particles that need to be examined. A large standard deviation results in a Gaussian filter with a large length which may affect the size of small particles, causing erroneous size measurements. The thresholds t_L and t_H, on the other hand, affect only the number of particles to be identified and not their size. High values lead to identification of fewer particles (only those with very sharp edges), while lower values increase the number of identified particles but also might lead to false positives (areas which happen to be surrounded by strong edges). Such areas correspond to fringes caused by conjugate image or out of focus particles and are much brighter than focused particles. As a result, they can easily be identified and ignored based on their average intensity.

The selection of these parameters is not critical for the performance of the algorithm and they do not have to be tuned for each hologram. This is verified by the results presented in Section 4.4, where the same set of parameters has been found to give acceptable results for all the experiments which cover a large number of holograms, recorded under different conditions and depicting different particle sizes.

Following edge detection, the dark areas that are completely enclosed by edges are filled into form blobs that are considered as possible particles. Open-ended lines are removed by erosion followed by dilation in order to eliminate noisy formations that frequently appear in the background. This does not change the size of the identified blobs. Blobs that are touching the edges of the reconstruction correspond to partially shown particles and are therefore removed in order to avoid size measurements corresponding to partial particles. Also blobs with very small diameter are removed as they are likely to correspond to noise.

Apart from good localization, the use of edge detection-based particle segmentation on digital holograms also has the advantage that particles are surrounded by strong edges only close to their best focusing point. As a result, highly unfocused particles are not considered to be easing depth localization. The procedure described above results in a set of blobs which correspond to particles shown in the reconstruction. These blobs have the same size and shape as the corresponding particles and can be used for particle measurement.

4.3.3 Particle Focusing

As can be seen from Equation 4.5, magnification depends on the object distance from the camera. On the other hand, it is practically impossible to know the exact location of the particles within the sample volume, hence this distance needs to be accurately determined from the recorded hologram for each particle separately. From each reconstruction, several properties of each identified particle such as its spatial location, its area in pixels, and a focusing metric are recorded. This results in a list of all the identified particles from several reconstruction depths. Each particle might appear several times in the list at different depths, but always at the same x–y location. Hence the multiple occurrences of each particle can be grouped together based on their location on the reconstruction. Each such group corresponds to a single particle, and the focusing metric is used to select the best focusing depth of the particle. Mean intensity and the variance of

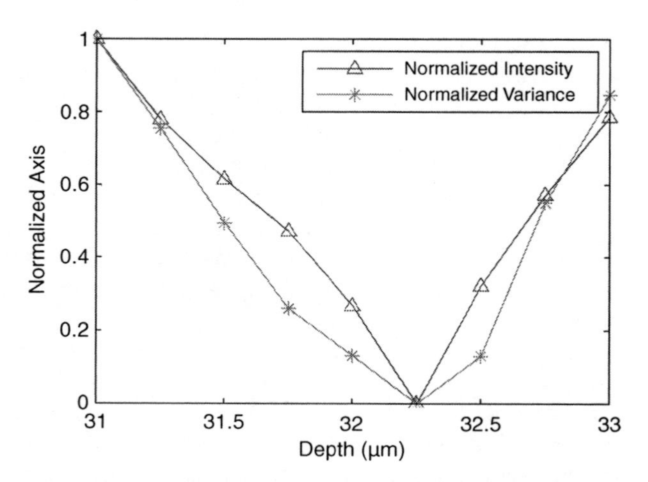

Figure 4.5 Depth profile of the normalized focusing metrics (mean intensity and variance of the intensity) for one particle. The best focusing depth for this particle is 32.25 mm

intensity have been used here as appropriate focusing methods [13]. The best focusing depth for each particle is selected as the depth where the focusing metric is minimized. The area of the particle at the best focusing depth is also selected as the best estimate of the particle's area. This procedure results in a list of uniquely identified and focused particles containing information such as their spatial location, depth and area.

Normalized depth profiles of the mean intensity and the variance of the intensity for one particle are shown in Figure 4.5. In this case, the best focusing depth is 32.25 mm. There are cases where the minimum of the particle's depth profile appears at the last or the first examined depth. In these cases, the particle is probably located outside the examined depth range and, hence, such particles are ignored to avoid erroneous measurements.

4.3.4 Particle Size Measurement

Following segmentation and particle focusing, the area which each particle occupies at its best focusing depth is determined accurately. In general, the particles are not spherical and, as a result, the identified regions are not circular. Accordingly, it is not always easy to measure their size. However, several properties of these areas can be extracted for further size and shape analysis. There are several ways to describe an object's size and shape, unless the object is perfectly spherical. The most commonly used factors describing these parameters are: circle equivalent diameter, aspect ratio, solidity and sphericity. These shape factors are illustrated in Figure 4.6. To make an appropriate choice

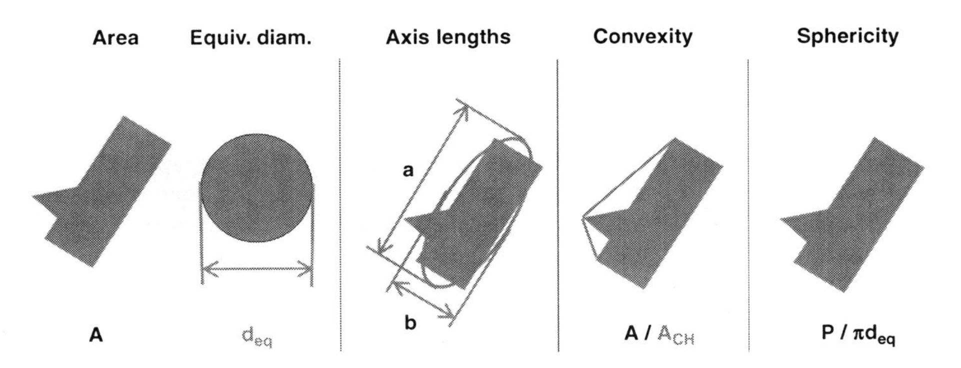

Figure 4.6 Shape factors

of shape factor, the definition of commonly used shape factors is discussed here briefly.

1. *Circle Equivalent Diameter.* Usually it is convenient to describe particle size as one single number which is most commonly calculated as circle equivalent (CE) diameter, that is, the diameter of a circle with the same projected area of the object. But the CE diameter is not always useful when obtaining unambiguous information about the particle shape, for example, very different shaped particles could be characterized as identical simply because they have similar equivalent areas.
2. *Aspect Ratio.* This is a measure of elongation of particles and is defined as major axis (a) to minor axis (b) length ratio (aspect ratio = a/b). Particles that are perfectly symmetrical in all axes, such as sphere or cube, will have an aspect ratio value of 1 and those with values greater than 1 will have an elongated shape. Hence, axis length distribution (ALD) provides 2-D size and shape information.
3. *Convexity.* This is the ratio of the area of the particle (A) to its convex hull area (ACH) (convexity = A/ACH) and, hence, is a measure of surface roughness of a particle. It is a useful selection criterion for single particles and agglomerates or fragmented particles. Generally, convexity for single particles is much larger than agglomerates or particle fragments.
4. *Sphericity.* This is a measure of deviation of a shape from a perfect circle. It is the ratio of the perimeter of the object (P) to the perimeter of a circle with the same projected area of the object (πd_{eq}) (sphericity = $P/\pi d_{eq}$). A perfect sphere has a sphericity of 1. Sphericity is sensitive to both overall form and surface roughness.

It should be noted that a single shape descriptor is not always sufficient to characterize different combinations of particle shapes. Among the aforesaid shape factors, aspect ratio has found to be more informative for 2D size and shape measurements. In this method both the equivalent diameter (the diameter of a circle with the same area as the identified regions) and the ellipses with the same normalized second central moment as the identified region can be used to extract the particle size distribution (PSD) or major axis and minor axis length distribution, respectively. In the case of equivalent diameter, the PSD will be correct in assuming that the particles are close to spherical shape. In the latter case, the identified ellipses characterize the particles by calculating the lengths of the major and minor axes. These lengths can be used to calculate the axis length distribution (ALD), or to differentiate between circular, ellipsoidal, and needle-like particles.

In order to convert the measured size to the real length, the value needs to be converted using the magnification factor M as:

$$r = \frac{r_{pixels}\Delta x}{M_{d_o}}$$

where r_{pixels} is the measured size of the particle in pixels, Δx is the pixel size of the recording camera, and M_{do} is the magnifying factor at the best focusing depth of each particle.

4.4 Validation and Performance of the 2D Measurement Method

In this section, several experiments are presented to verify the performance of the algorithm described in Section 4.3.1. The general set-up shown in Figure 4.2 was used with a green laser of wavelength, $\lambda = 532$ nm, a 60X microscope objective lens and 1 μm pinhole. The distance between the point source and the CCD camera for recording was $D = 62$ mm, and for the reconstruction a point source to CCD camera distance $D' = 100 \times D = 6.2$ m was used. The camera used for the experiments had 1280×960 square pixels of size $\Delta x = \Delta y = 4.65$ μm. Prior to all experiments the set-up was tested with a USAF target. The obtained resolution was ≈7 μm for $D = 62$ mm.

The threshold values and standard deviation of the Gaussian filter that were used for the edge detection algorithm for all the experiments were chosen as $t_L = 0.3$, $t_H = 0.6$ and $\sigma = 1.5$, respectively. Sigma, $\sigma = 1.5$ results in a filter size of 6×6 pixels. This filter length has been found to give good noise reduction without affecting the size of the smallest particles studied in these experiments which had a diameter of ~40 μm or ~17 pixel particles. These parameters were retained for the first two experimental systems listed in Table 4.1. In the third experimental system, particles with a diameter of 10 μm or ~3 pixels were studied. Due to the small size of these particles, no Gaussian

Table 4.1 Characteristics of various particles used in the study

Experimental system	Average size	Shape	Transparency
Ceramic beads on glass slide	Diameter ≈ 80 μm	Spherical to elliptical	Opaque
Polymer particles in flow system	Diameter = 40 μm	Spherical	Opaque
Microsphere suspension	Diameter = 10 μm	Spherical	Opaque
Single fiber on glass slide	Length = 1320 and 150 μm	Needle	Opaque
Carbon fibers in suspensions	Diameter ≈ 9 μm, length ≈ 50–500 μm	Needle	Opaque

filtering was used prior to edge detection, and in the fourth and fifth experiments, carbon fibers of 9 μm diameter and varying length were used, hence, the Gaussian filter size was reduced to $\sigma = 1$.

4.4.1 Verification of the Focusing Algorithm

This section describes an experiment that is used to verify the accuracy of the focusing step. Section 4.3.1 presents other experiments which verify the performance and the accuracy of the whole particle measurement procedure. In order to verify the accuracy of the focusing method, a series of experiments were carried out. For these experiments, the object consisted of seven ceramic beads with average size of ~80 μm positioned on a glass slide. The slide was positioned in the place of the object shown in Figure 4.2, normal to the optical axis so that the particles were located at the same depth. A hologram of this slide was captured. The slide was then consecutively displaced by 1 mm along the optical axis for number of times, and a hologram was recorded for each displacement. The experiment was then repeated for a slide displacement of 0.1 mm. In both cases, the displacement was achieved with a device of accuracy ±0.1 mm. Following this, the focusing algorithm described in Section 4.3.1 was used to find the best focusing point for the particles of each hologram. In the case of 1 mm displacement, the algorithm was used to find the best focusing point within a depth range of 27.5–32.5 mm and in the case of 0.1 mm within a range of 20–34 mm. In both cases, the depth step was selected as 50 μm. Following the focusing of the particles, the depth of the slide for each position was estimated by averaging the depth of the particles. Table 4.2 shows the measured slide positions. The best focusing depth of the particles and the estimated slide positions are shown in Figure 4.7a for the displacement of 1 mm and in Figure 4.7b for the displacement of 0.1 mm.

Table 4.2 Numerical results of the focusing algorithm for displacement of (a) 1 mm and (b) 0.1 mm

d (mm)	Measured displacement (mm)	Expected displacement (mm)	Error (mm)
(a)			
32.26	—	—	—
31.19	1.07	1	0.07
30.11	1.08	1	0.08
29.03	1.08	1	0.08
27.91	1.12	1	0.12
(b)			
32.14	—	—	—
32.06	0.08	0.1	−0.02
31.94	0.12	0.1	0.02
31.81	0.13	0.1	0.03
31.73	0.08	0.1	−0.02

Source: 'Processing of digital holograms for size measurements of microparticles' by Emmanouil Darakis, Taslima Khanam, Arvind Rajendran, Vinay Kariwala, Anand K. Asundi, Thomas J. Naughton. Published in Proceedings of SPIE, 2008

According to Table 4.2, the maximum observed error between the measured and the expected slide displacement is 120 μm. Also, assuming that all the particles are located at the same depth as the slide, the maximum observed error between a particle's depth and the corresponding slice depth is 110 μm, as can be seen in Figure 4.7. According to Equation 4.5, a 110 μm focusing error causes an error in magnification of ~0.35% under these conditions.

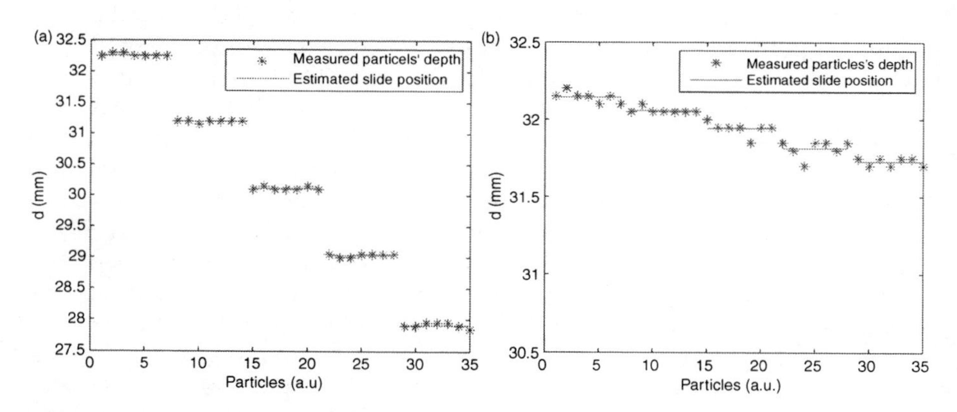

Figure 4.7 Verification of the focusing algorithm. (a) For 1 mm displacement; (b) For 0.1 mm displacement
Note: Asterisks show the measured position of the particles and the lines show the estimated slide depth (average depth of the particles)

4.4.2 Spherical Beads on a Glass Slide

In order to verify the performance of the particle measurement algorithm, ceramic beads from a population of average bead sizes ≈80 μm were used. Several holograms of particles from the population positioned on glass slides were recorded using the set-up shown in Figure 4.2. The recorded holograms were processed following the procedure described in Section 4.3. A depth of 3 mm with a step size of 50 μm was used for each hologram. The algorithm identified 437 different particles.

In addition, Scanning election microscope (SEM) was used to record several images of different particles taken from the same population. One such image is shown in Figure 4.8a. The SEM images were segmented using Canny edge detection, using the same parameters as before, to extract the particles (bright areas). The size of each particle was measured considering the magnification of the SEM. 615 different particles were measured from the SEM captured images.

Figure 4.8b shows the resulting distributions of equivalent diameters of ceramic beads obtained from the holography and SEM experiments. The mean particle size identified from the holographic microscopy and SEM are 80.24 ± 14.46 μm (mean ± standard deviation) and 79.23 ± 13.79 μm, respectively, a good agreement indeed. In the case of SEM, focusing was performed manually during the recording, and the magnification was provided automatically from the instrument. On the other hand, in the case of digital holography, focusing and calculation of the magnification factor were performed numerically. Hence, this experiment verified the accuracy of digital holography based measurement algorithm.

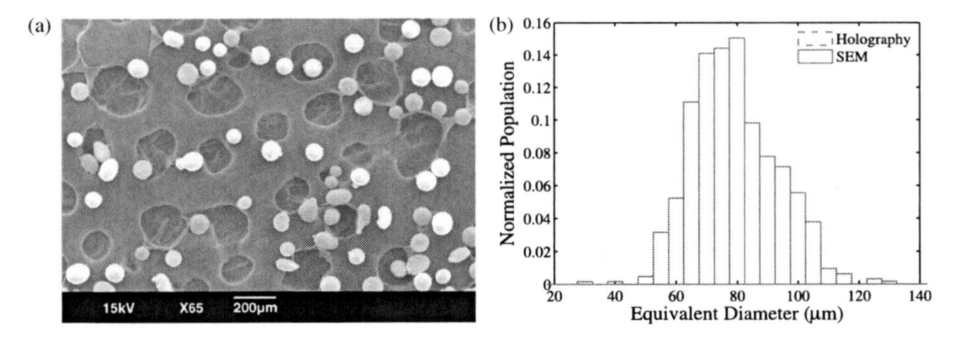

Figure 4.8 (a) An image of ceramic beads obtained from SEM. Bright areas correspond to particles whereas dark circular areas on the background are irregularities of the sample holder. (b) Comparison of PSD obtained from digital holography and SEM (Reproduced by permission of © 2009 Elsevier)

4.4.3 Microspheres in a Flowing System

For this experiment, the object used in the set-up consisted of a flow cell with flowing particle suspensions. National Institute of Standards and Technology (NIST) certified polymer microspheres manufactured by Duke Scientific Corporation, USA, with a diameter of $40.25 \pm 0.32\,\mu m$ were continuously pumped from a beaker through a circulation loop using a peristaltic pump. The particles flowed through a flow cell equipped with quartz windows which enabled the imaging. Holograms of the flowing particle suspensions through a 12.5 mm (L) $\times$ 12.5 mm (W) $\times$ 65 mm (H) flow cell with an optical path length of 10 mm at a flow rate of 5ml/min were obtained. Holograms were captured at the rate of one hologram per second. Reconstructions were carried out with a distance of $25\,\mu m$ between each one, covering a volume of depth size of 8 mm.

Figure 4.9a shows one of the recorded holograms of $40\,\mu m$ polymer particles' suspension in the flow cell and the corresponding reconstructed

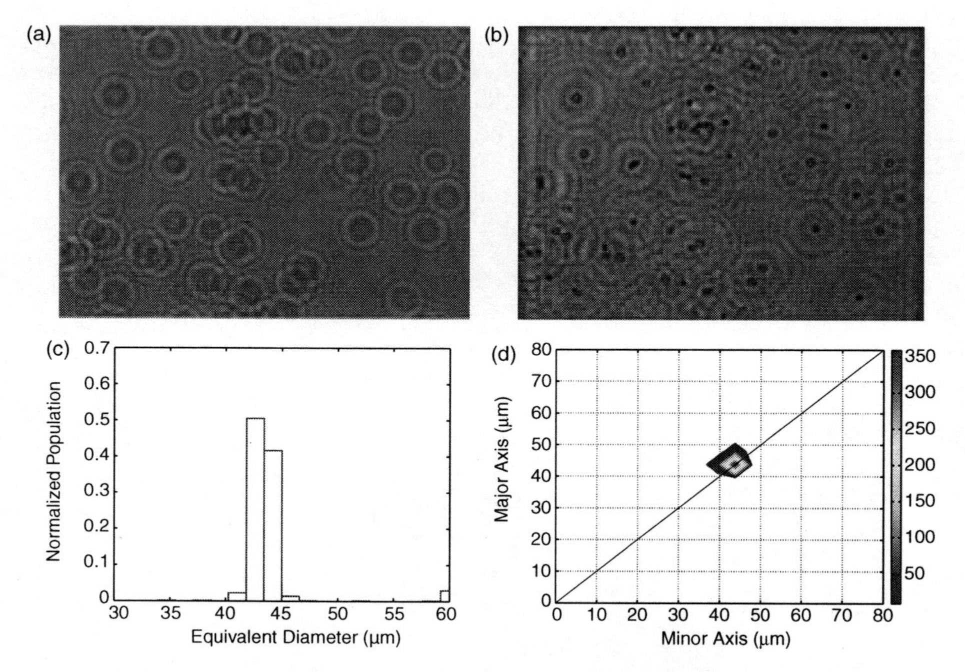

Figure 4.9 Size measurements of polymer particles in flow-through system. (a) Hologram; (b) Sample reconstruction; (c) Distribution of equivalent diameter. Obtained mean particle size is $43.89 \pm 3.38\,\mu m$ whereas expected mean diameter is $40.25 \pm 0.32\,\mu m$; (Reproduced by permission of © 2009 Elsevier) (d) Distribution of the axis lengths representing spherical shape of particles

image is shown in Figure 4.9b. The algorithm identified 437 different particles. The resulting particle size distribution (PSD) and axis length distribution (ALD) are shown in Figures 4.9c and d, respectively. The obtained mean particle size was $43.89 \pm 3.38\,\mu m$. There is an error of $\approx 4\,\mu m$ between the actual particle size and the location of the PSD peak. This error is below the resolution of the system (the obtained resolution for $D \approx 62\,mm$ was $\approx 7\,\mu m$). In Figure 4.9c, a spread in PSD $\approx 60\,\mu m$ is also observed. This indicates the presence of agglomerated particles. The ALD shown in Figure 4.9d is indicative of the shape of the particles and can be used to classify the shape of the particles. As can be seen, the obtained ALD in Figure 4.9d lies on the diagonal line confirming the spherical shape of the particles.

4.4.4 10 μm Microspheres Suspension

In this experiment, 10 μm particles suspended in water were examined and 40 reconstructions with a distance of 25 μm between each were obtained from one hologram to cover a depth of 1 mm. Accordingly, 47 different particles were identified from the algorithm. Figure 4.10a shows one reconstruction of the hologram and Figure 4.10b shows the calculated PSD. The particle size is very close to the resolution limit of the system ($\sim 7\,\mu m$) leading to the relatively large spread around the expected size which can be seen in Figure 4.10b.

4.4.5 Measurement of Microfibers

Unlike spheres, measurement of fibers deserves special attention as its orientation in 3D volume affects the measurement. It is worth noting that only the projection of the object appears on the 2D camera plane. Hence, in

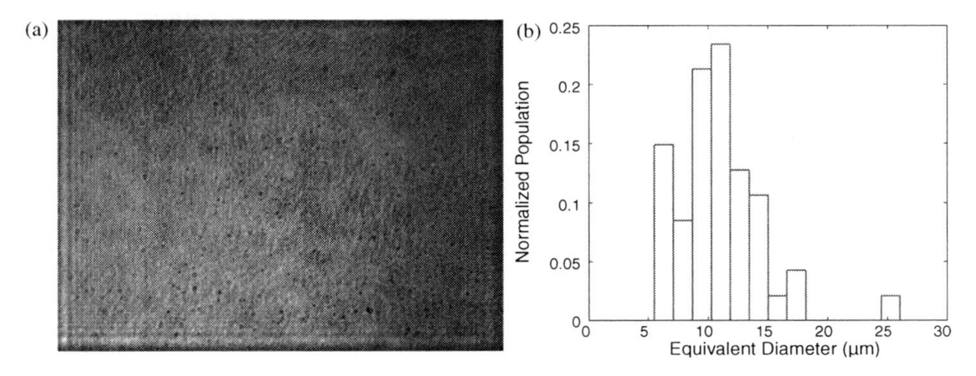

Figure 4.10 Digital hologram of 10 μm particles suspended in water. (a) Example of reconstruction; (b) Measured PSD from 47 identified particles

the case of spheres, regardless of its orientation in a 3D volume, the projection always provides the complete diameter of the sphere. But for needle-shaped particles, the potential to be tilted out of the camera plane are high. As a consequence, only a part of the fiber is focused in any reconstruction. This leads to the underestimation of the projected length of the fiber-like particles. If the fiber length is so short that it is comparable to the depth of focus, then only the entire projection of the fiber might occur on a single reconstruction. Under these circumstances, reconstructing at that particular distance yields either the underestimated projected length or the entire projected length of the fiber. Hence, from the 2D projection-based method one cannot discover the real length of the fiber. In order to study this, two sets of experiments were performed. The first set involved the measurement of a single fiber placed on a glass slide, allowing it to be rotated at a different out-of-plane tilt. In the second set of experiments, measurements of a population of fibers in a solution are reported. In these experiments, particles with a needle shape were studied using carbon fibers (TOHO Tenax Type 383) suspended in water.

4.4.5.1 Single Fiber on a Glass Slide

In the first set of experiments, a single fiber was placed on a glass slide which is connected to a rotatable mount. The mount can be adjusted to obtain the desired out-of-plane tilt with 1° accuracy. Two fibers with lengths 150 and 1320 μm, respectively, were allowed to rotate from 0° to 70° tilt at 5° intervals with respect to the optical axis (Z). The hologram of each rotation is captured. The holograms were processed according to the algorithm discussed in Section 4.3. Since the fibers are clearly non-spherical, elongation of the fiber is obtained by fitting an ellipse around it. This provides the major and minor axis lengths of the fiber. As the fibers are rectangular rather than elliptical, fitting an ellipse leads to an overestimation of the length of its axes. Hence, the measured axes lengths are multiplied by a factor of $\sqrt{3}/2$ to convert them to the fiber length and diameter accordingly [4].

The comparison between the measured and expected projected lengths for both the short and the long fiber is shown in Figures 4.11a and b, respectively. As it can be seen, the algorithm successfully estimated the projected length of the 150 μm long fiber within the expected accuracy even for large out-of-plane tilts. For the 1320 μm long fiber, the algorithm measured the fiber only for small tilts (Figure 4.11b). For higher tilts, long fibers are not entirely focused on a single reconstruction and hence the algorithm ignores such fibers, as they are not enclosed by strong edges. In this way, erroneous measurements are avoided but long fibers with significant off-axis tilt are preferentially ignored,

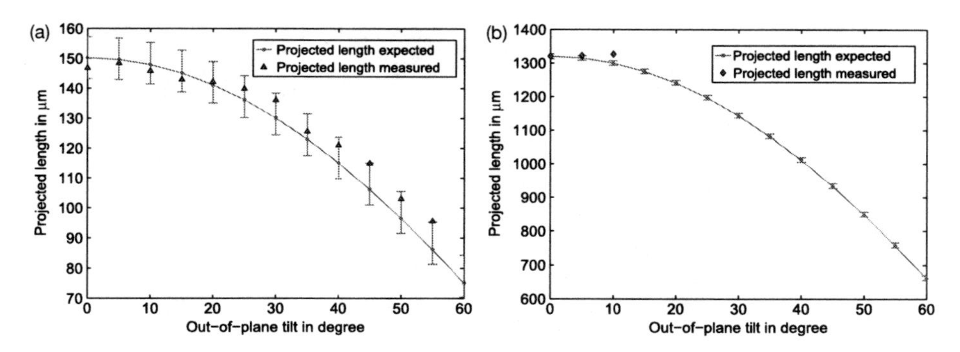

Figure 4.11 Measurements of fibers for out-of-plane tilt using 2D measurement method. (a) 150 μm long fiber; (b) 1320 μm long fiber (Reproduced by permission of © 2009 Elsevier)
Note: The error bars take into account the system resolution and the uncertainty in the measurement of the tilt

leading to false negative. It is worth pointing out that in the case of fibers, the measured lengths that are eventually identified correspond not to the real lengths but only the projected lengths. The measurement of the real length will be discussed in the 3D measurement section.

4.4.5.2 Carbon Fibers Suspended in Water

Based on the understanding obtained from the single fiber tilt experiments, a population of carbon fibers in suspension is studied. A series of holograms of the population of fibers contained in a quartz cuvette with dimensions 12.5 mm (L) × 12.5 mm (W) × 48 mm (H) with an optical path length of 10 mm were recorded and reconstructed with a step size of 20 μm between two successive reconstructions covering the volume depth of 8mm. One of the recorded holograms and, a sample reconstruction are shown in Figures 4.12a and b, respectively. Using the procedure described in Section 4.3, 13 holograms were analyzed and it identified 283 fibers. The resulting major and minor axes length distributions are shown in Figures 4.12c and d, respectively. The obtained mean major and minor axes length of these fibers were 138.74 and 13.34 μm, respectively.

Figure 4.12e is the representation of the axis length distribution. While spherical particles tend to lie close to the diagonal, this ALD tends to lie near the ordinate axis due to their high major to minor axis ratio. This clearly confirms that the population contains needle-shaped particles with similar widths but varying lengths.

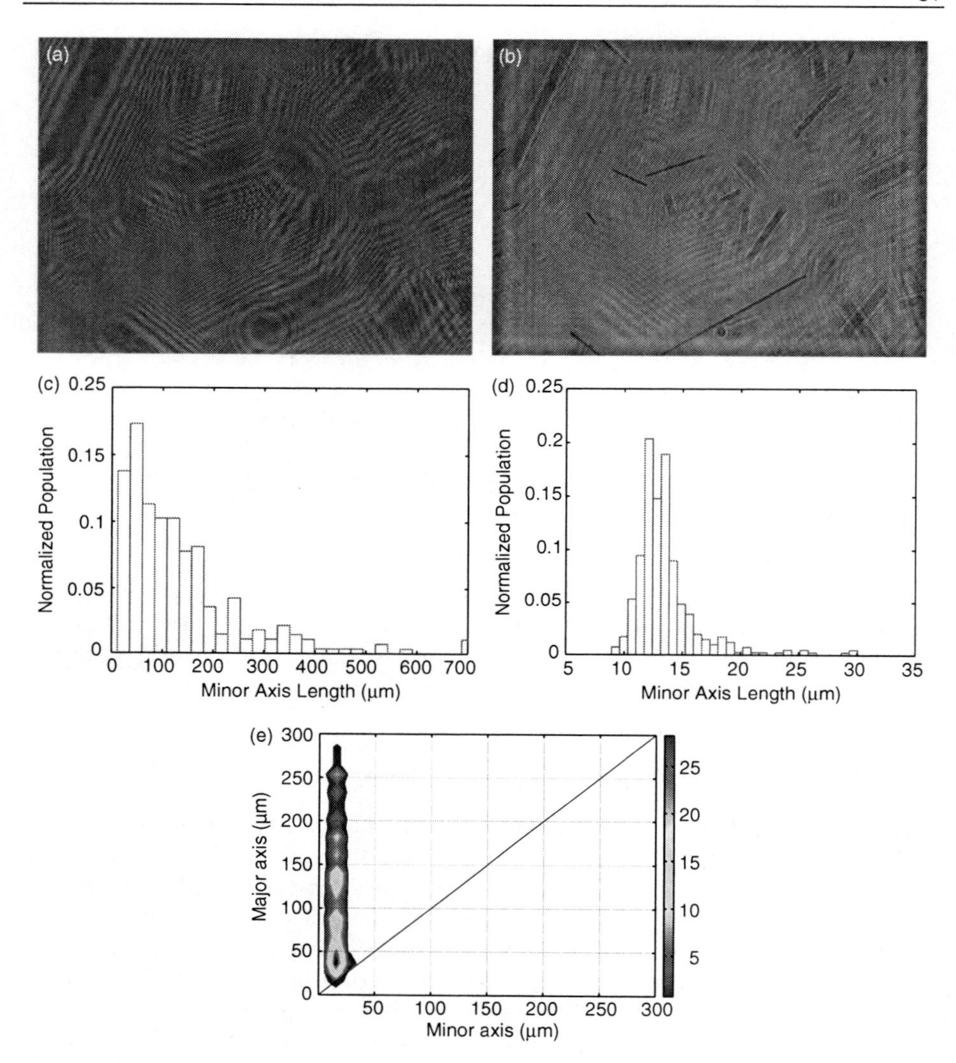

Figure 4.12 Size measurements of carbon fibers suspended in water. (a) Hologram; (Reproduced by permission of © 2009 Elsevier) (b) Sample reconstruction; (Reproduced by permission of © 2009 Elsevier) (c) Major axis length distribution; (Reproduced by permission of © 2009 Elsevier) (d) Minor axis length distribution; (e) Axis length distribution representing the needle-like shape of the particles (Reproduced by permission of © 2009 Elsevier)

4.5 Methodology for 3D Measurement of Micro-Fibers

The measurement of the particles in the plane of the camera limits the applications of the system for highly oriented fibers in 3D volume. This understanding has already been gained in single fiber experiment in Section 4.4.5. In order to overcome the limitation associated with tilted fibers, two techniques to measure the real lengths of fibers from hologram reconstructions directly without *a priori* knowledge of fiber tilt, position and length are presented here.

4.5.1 Method 1: The 3D Point Cloud Method

This algorithm consists of four steps: recording and reconstruction, thresholding, 3D reconstruction and particle characterization [10]. Here, in the first step, the digital hologram of the sample is recorded and numerically reconstructed at several depths, with $0 < d < D$, covering the volume of the sample to be studied using the method described in Section 4.2.1. The number of reconstruction steps is typically set by the minimum shortest length of fiber that is expected to be detected. Small values of d result in increasing the total number of reconstructions, thereby resulting in increasing the computational load. Larger increments in between two successive reconstructions might lead to errors in determining the sizes, as some focused part of the object may have been missed out in that depth interval. Hence the number of reconstructions should be set close to the minimum shortest length of the fiber.

4.5.1.1 Thresholding

After reconstructing the hologram at Nz equally spaced intervals, histogram equalization is performed on the intensity of complex valued reconstructions to account for variations of the intensity's range. The gray scale images are then converted to binary images using a simple threshold. The selection of the threshold is such that only the darkest pixels that correspond to pixels in focus are accepted as being part of the fiber and are set to 1 in the binary image. The rest of the pixels are considered to be background and their value is set to 0 in the binary image. Here, a threshold of 5 (on a scale of 0 (black) to 255 (white)) was used for the equalized images and almost identical results were obtained for threshold values up to 15. A sample hologram of a tilted fiber and the corresponding reconstructed image at a particular depth and the resulting thresholded image are shown in Figures 4.13a–c, respectively. In Figure 4.13c, in-focus parts of the fiber appear in black. The binary images obtained after thresholding are stored in a 3D matrix A of size $Nx \times Ny \times Nz$, whereas $Nx \times Ny$ is the dimension of the image plane and Nz is the number of reconstruction. Thus the matrix A contains different clouds of points of in-focus pixels of a fiber at different reconstruction depths. Based on the connectivity of the pixels in 3D, identical group of pixel clouds are assigned and labeled. Hence each group of point clouds corresponds to a different fiber. Now there might possibly appear broken up fibers due to the high threshold. In order to merge the broken up parts, 3D dilation followed by 3D erosion is performed on A. After this, a cut-off value equivalent to a minimum number

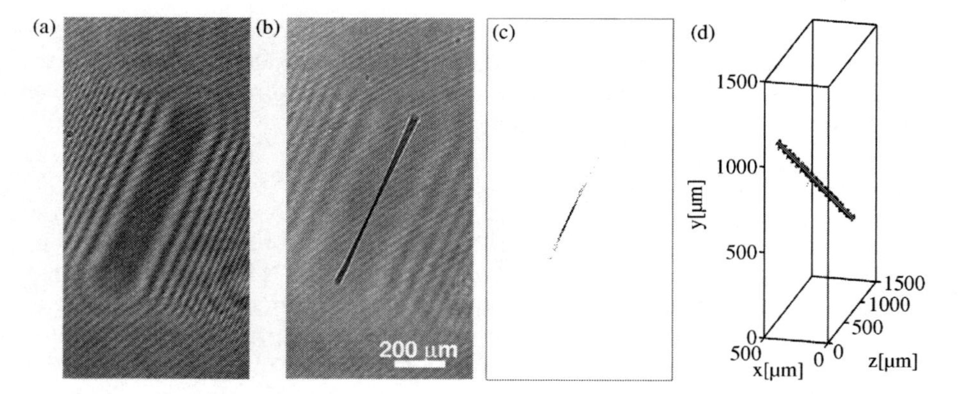

Figure 4.13 3D point cloud method for measuring tilted fiber. (a) Hologram; (b) A sample reconstruction; (c) Threshold reconstruction; (d) Point cloud with fitted lines

of connected pixels is set to remove any unwanted noise. Furthermore, any group of connected pixels touching the reconstruction edges is also removed, as they are likely to be part of a fiber partially located outside the image frame. Following this, connected pixels within A are grouped together. Figure 4.13d shows the point cloud of the fiber in black.

4.5.1.2 Size and Orientation Measurement by 3D Line Fitting

In this step, each group of pixels corresponding to individual fibers is fitted with a straight line to yield their orientation and length. As can be seen in Figure 4.13b, the intensity of the reconstruction is contaminated by speckle noise. This results in noisy point clouds that can be seen in Figure 4.13d. Traditional regression methods, such as least squares, are based on the assumption of error-free predictors and are thereby inadequate for this situation. Hence, tools like principal component analysis (PCA) has been utilized for the fitting line in the point clouds, since it can account for errors in the predictors. Figure 4.13d shows the line in black that was fitted in the point clouds of a tilted fiber. From this line, the characteristics of the corresponding fiber (location, size, and orientation) can be measured. Results verifying the accuracy of the proposed method are discussed in Section 4.6.

4.5.2 Method 2: The Superimposition Method

Since the 3D point cloud method requires computation of a large 3D matrix, this method is highly memory-intensive. In order to overcome this problem an alternate method of the 3D point cloud, namely the superposition method, is introduced here [14]. The superposition algorithm mainly consists of the following steps: (1) recording and reconstruction; (2) image segmentation; (3) superimposition; and (4) localization of randomly oriented fiber. The first step is identical to the procedure described in Section 4.2.1.

4.5.2.1 Image Segmentation

Similar to the first step of the 3D point cloud method, *Nz* numbers of reconstructions of a hologram are checked for threshold value after performing histogram equalization to account for variations in the intensity. In this method, the selection of the threshold is relative, based on the quality of the hologram, as variation in the intensity due to laser imperfections, multiple object light scattering and issues relating to the non-uniform illumination, has to be considered. The threshold value for each hologram is selected as a percentage of average intensity of a reconstruction. Here, 10% and 20% of the mean intensity are used as threshold values for the analysis of the holograms of dry particles and suspended particles, respectively. After applying the threshold, the intensity image is converted to a binary image where 1 valued (white) pixels are for in-focus pixels and 0 valued pixels (darkest) are considered the background.

4.5.2.2 Superimposition

In this step, each binary reconstruction is superimposed onto the preceding image until the reconstruction of the whole sample volume is covered (see Figures 4.14a–d). In this way, all the in-focus pixels at a different depth that correspond to a fiber are merged together, comprising a full projection of the fiber on a single 2D image plane (see Figure 4.14e). After superimposition, the resulting image became a double class image which is again converted to a binary image. Following this, a morphological operation, namely bridging, is performed on the superimposed binary image to join the 0 valued pixel gap between two non-zero neighbors. After bridging, dilation, followed by erosion, is performed to merge the rest of the broken pieces of fiber that are likely to be

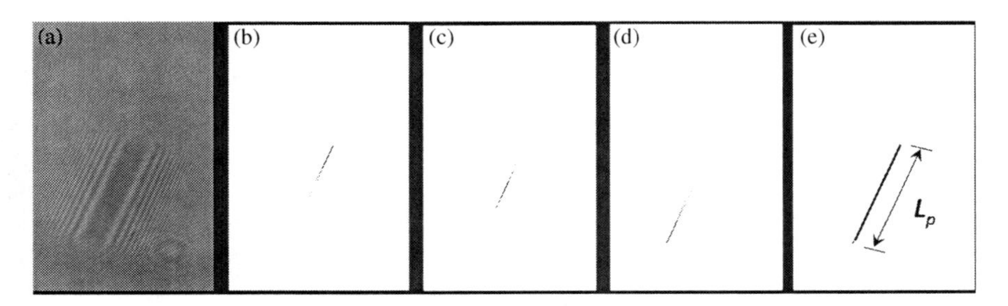

Figure 4.14 Superimposition method for measuring tilted fiber. (a) Hologram; (b–d) Threshold reconstructions at different depths; (e) Resulting superimposed image

parts of fiber. In addition, a tailor-made co-linearity method, where morphological operations are not sufficient, is used. This co-linearity method finds the fiber pieces which are parallel to each other and checks the distance between two vertices to merge the rest of the broken pieces that might correspond to a part of complete fiber. Hence, from the superimposed image, the projected length of the individual fiber is obtained. It is worth pointing out, unlike the 3D point cloud method, that all image processing operations are performed on a 2D image plane, thereby saving the memory and computational load. Finally, all the detected fibers are labeled and the projected length is measured from the major axis length of the fitted ellipse around the fiber on the 2D image plane. Hence the superimposition method provides the complete projected length of a fiber in a 2D plane.

4.5.2.3 Localization of a Tilted Fiber

The superimposition step yields the complete projected length of a fiber despite a large out-of-plane tilt. In order to know the real length of a fiber, its orientation needs to be measured. The localization step identifies the position of the tilted fiber in the 3D volume. For this purpose, the algorithm selects two end regions by sectioning the projected length of the individual fiber into four parts. Thus, the first and fourth sections are considered as the two end sections. The mean intensity of the two selected regions is calculated for all reconstruction depths within the volumes, see Figure 4.15a. Hence, the two selected end regions are expected to have the minimum mean intensity at their best focused planes.

The depth profiling to find the focused positions of the two end regions of a single tilted fiber is shown in Figure 4.15b. Once the projected length and two axial positions of a tilted fiber are known, the real length is calculated as:

$$\theta = \tan^{-1}\frac{\Delta d}{L_p} \tag{4.7}$$

$$L_R = \frac{L_p}{\cos\theta} \tag{4.8}$$

where θ is the angle of orientation along the optical axis, Δd is the distance between the two positions of a tilted fiber along the optical axis (axis Z in Figure 4.2), and L_P, L_R are the projected and real length, respectively. As can be seen from Equations 4.7 and 4.8, the measured θ depends on L_P and Δd, which affects the measurement in the real length. It is worth pointing out that if the projected length of a fiber is comparable to the system depth of focus, then it affects the measurement of θ significantly, as the two end positions of the tilted fiber are not clearly distinguishable due to the system depth of focus. This is particularly observed at the higher degree tilt, as the projected length becomes

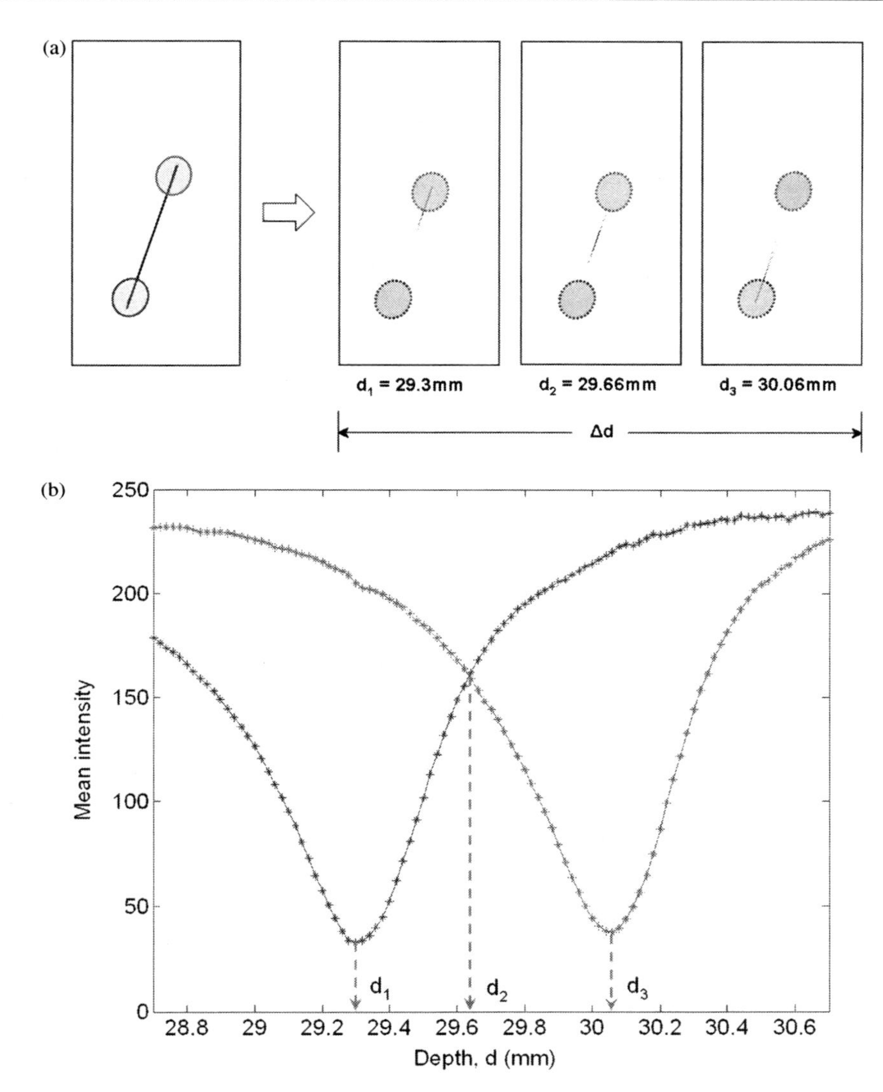

Figure 4.15 Localization of a tilted fiber: (a) Terminal focusing areas for calculating average intensity; (b) Intensity profile of the terminal focus areas determines the tilted position of the fiber

smaller with the increase in out-of-plane tilt. Hence, the measurement for a highly tilted fiber is not reliable. To avoid this, the algorithm identifies the highly tilted fibers and discards them from the measurement. The maximum number of connected in-focus pixels among different reconstruction depths that correspond to a fiber segments can be a measure of high degree tilt.

Low tilted fibers are expected to have at least one fiber segment of a size larger than 70 pixels at any reconstruction depth. If any segment of a fiber does not have a size larger than 70 pixels, at least in any one of the reconstruction planes, the fiber is considered to be highly tilted. Thus, the algorithm successfully identifies the highly tilted fibers and discards them from the measurement. It should be noted that this cut-off value, 70 pixels, becomes more stringent for short fibers. The shorter the length of the fiber, the smaller the value of the tilt would be considered a high degree tilt.

4.6 Validation and Performance of the 3D Measurement Methods

In this section, the performance of the 3D point cloud algorithm and the superimposition method is analyzed. Furthermore, the more efficient method will be demonstrated for microfiber suspension in comparison with microscopy measurements. In order to verify the algorithm performance, the single fiber tilt experiment for long and short fibers described in Section 4.4.5 is used.

4.6.1 Experiment with a Single Fiber

Figure 4.16 shows the relative error in the measured length for the same long and short fibers for different out-of-plane tilts using the superimposition method and the 3D point cloud algorithm. It should be noted that, for the sake of comparison of the two methods, the results using the superimposition method show all the measurements without ignoring the high degree tilt as was ignored using the method in Section 4.5.2.3. As can be seen from Figure 4.16a, the superimposition method accurately measures the fiber length,

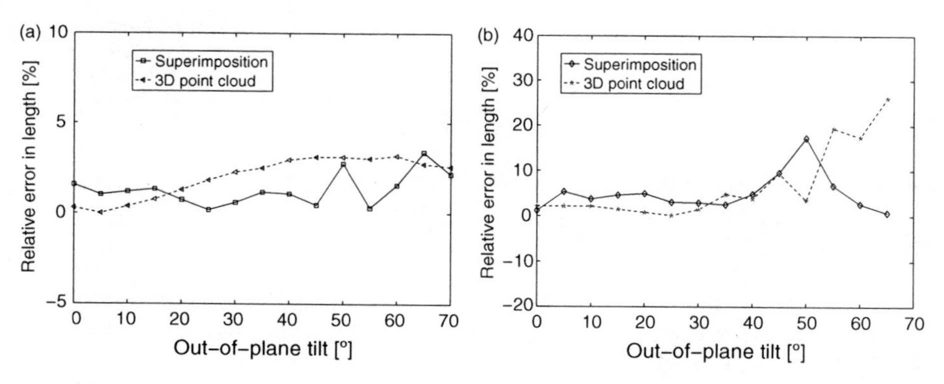

Figure 4.16 Measurement of real length of a single tilted fiber. (a) Long fiber (1320 μm); (b) Short fiber (150 μm)

which corresponds very well to the true length of the fiber over a large range of tilts with a relative error less than 4%. The measurement accuracy for both 3D point cloud algorithm and superimposition method are comparable for the case of long fiber, while the latter method comes with the increased advantage of being simpler and faster than the former.

Figure 4.16b shows the results from the measurement of the short fiber. As can be seen from Figure 4.16b, both the superimposition and the 3D point cloud methods are comparable and at lower degree tilts, the measurement error is in the range of 5% in both cases. But with a tilt of more than 40°, the measurement error seems to be increasing in both methods. This comes from the limit of the system's depth of focus. However, the superimposition method successfully ignores the measurements after the 55° tilt and the 35° tilt for 1320 and 150 μm fibers, respectively. To enhance the applicability of the superimposition method for fibers less than 150 micron dropping down to near a micron, the depth resolution of the system must be improved which can be achieved by the use of microscope objective [15].

In the following section, only the superimposition method is demonstrated for the population of micro-fibers due to some of its added advantage, such as ignoring the measurement of high degree tilt, and less memory-intensive use.

4.6.2 3D Measurements of Micro-Fibers in Suspension

After the validation steps of the two methods, the superimposition method was applied to the population of fibers, a situation encountered in practice. For these experiments, the rotating mount was replaced with a cuvette 12.5 mm (L) × 12.5 mm (W) × 48 mm (H) containing a suspension of a population of carbon fibers (Asbury PAN type carbon fiber, USA) in water. For validation purposes, true lengths are measured with microscopy measurements. For microscopy measurement, carbon fibers were placed on glass slides and their lengths were measured using a microscope. Over 600 microscopy images were acquired and the lengths of 12 258 fibers were measured by image analysis. A total of 514 holograms of fiber suspension were recorded from which 1668 fibers were identified.

A sample hologram, a corresponding superimposed image after segmentation and the corresponding 3D distribution of the identified fibers are shown in Figures 4.17a–c, respectively. The resulting real length distribution of fiber population obtained by analyzing 514 holograms is shown in Figure 4.17d. In Figure 4.17d, distribution of real length obtained by analyzing 600 microscopy images of the same fiber population has also been compared with the holography measurement. It should be noted that Figure 4.17d is plotted for fibers that were sorted to the lengths longer than 150 micron, as the robustness of the method for fibers less than 150 micron is poor, due to the depth of focus limit of the system. Hence, in order to compare them with the holography data,

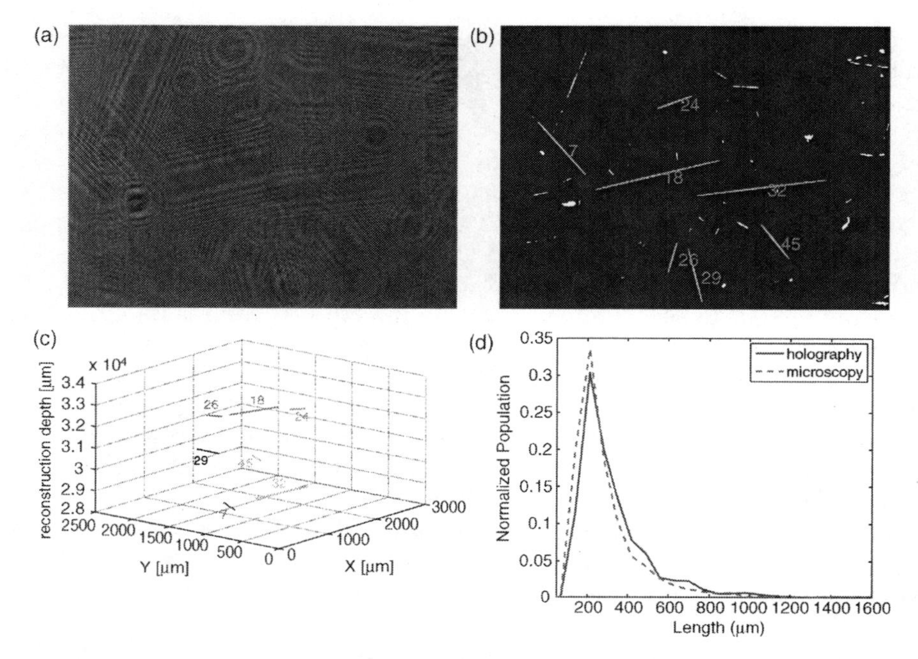

Figure 4.17 (a) A sample hologram of fiber suspension in a cuvette. (b) Superimposed image of all reconstructions of the hologram; (c) 3D distribution of identified fibers showing their orientations and lengths in volume; (d) Real length distribution of the fiber population longer than 150 μm in suspension

microscopy measurements are also sorted for fibers longer than 150 micron. As can be seen from Figure 4.17d, both distributions match very well, attesting to the accuracy of the method for fiber suspension.

4.7 Conclusion

This chapter has discussed the application of digital in-line holography for particle size and shape measurement. Automated methods for processing digital holograms, subsequent image analysis and the necessary computation for measuring various particle sizes and shapes have been discussed. Issues related to the orientation of particles in a 3D volume and the impact of their shape on it have been highlighted, and appropriate methods to solve the issues have also been shown and benchmarked.

From the study of this chapter, it can be concluded that digital in-line holography is a powerful tool to measure particle size and shape. It has the capability to retrieve the real length of randomly oriented needle-like particles from

3D volume directly without any measurement model. Thus it opens up avenues for 3D measurements of particles where simultaneous measurement of size, position and angle of orientation is of great interest. It should be noted that the in-line digital holography set-up studied in this chapter is transmission-based. Hence, care should be taken when handling suspension density to obtain the best results. The higher the suspension density, the less intense the reference wave, thereby the quality of the hologram will be degraded. Another limitation of the lens-less set-up is the limited magnification, as the magnification depends only on the system geometry. Hence, in order to study near micron particles, a microscope objective needs to be used where appropriate imaging equations need to be modified accordingly.

All the particles studied in this chapter are of opaque characteristics, but it is worth pointing out that the methods discussed here also have the potential to handle transparent particles. Despite this, more investigation and research need to be conducted to establish the best method for measuring transparent particles appropriately.

References

1. Larsen, P.A. and Rawlings, J.B. (2009) The potential of current high-resolution imaging-based particle size distribution measurements for crystallization monitoring. *AIChE J.*, **55** (4), 896–905.
2. Hukkanen, E.J. and Braatz, R.D. (2003) Measurement of particle size distribution in suspension polymerization using in situ laser backscattering. *Sensor Actuat. B-Chem.*, **96** (1–2), 451–459.
3. Vikram, C.S. (1992) *Particle Field Holography*, Cambridge University Press, Cambridge, UK.
4. Darakis, E., Khanam, T., Rajendran, A. *et al.* (2010) Microparticle characterization using digital holography. *Chem. Eng. Sci.*, **65** (2), 1037–1044.
5. Darakis, E., Khanam, T., Rajendran, A. *et al.* (2008) Processing of digital holograms for size measurements of microparticles. *Proc. SPIE Int. Soc. Opt. Eng.*, **7155**, 715524-12.
6. Khanam, T., Darakis, E., Rajendran, A. *et al.* (2008) On-line digital holographic measurement of size and shape of microparticles for crystallization processes. *Proc. SPIE Int. Soc. Opt. Eng.*, **7155**, 71551k-10.
7. Xu, W., Jericho, M.H., Meinertzhagen, I.A., and Kreuzer, H.J. (2002) Digital in-line holography of microspheres. *Appl. Opt.*, **41**, 5367–5375.
8. Kempkes, M., Darakis, E., Khanam, T. *et al.* (2009) Three dimensional digital holographic profiling of micro-fibers. *Opt. Express*, **17** (4), 2938–2943.
9. Xu, L., Peng, X., Guo, Z., Miao, J., and Asundi, A. (2005) Imaging analysis of digital holography. *Opt. Exp.*, **13** (7), 2444–2452.
10. Schnars, U. and Juptner, W. (2005) *Digital Holography: Digital Hologram Recording, Numerical Reconstruction, and Related Techniques*, Springer, Berlin.
11. Asundi, A. and Singh, V.R. (2006) Circle of holography—digital in-line holography for imaging. *J. Holography Speckle*, **3**, 106–111.

12. Canny, J. (1986) A computational approach to edge detection. *IEEE T. Pattern Anal.*, **8**, 679–698.
13. McElhinney, C.P., Hennelly, B.M., and Naughton, T.J. (2008) Extended focused imaging for digital holograms of macroscopic three-dimensional objects. *Appl. Opt.*, **47**, D71–D79.
14. Khanam, T., Rahman, N.R., Rajendran, A., Kariwala, V., and Asundi, A.K. (2011). Accurate size measurement of needle-shaped particles using digital holography *Chem. Eng. Science*, doi: 101016/j.ces.2011.03.026.
15. Sheng, J., Malkiel, E., and Katz, J. (2006) Digital holographic microscope for measuring three-dimensional particle distributions and motions. *Appl. Optics*, **45** (16), 3893–3901.

5

Other Applications

5.1

Recording Plane Division Multiplexing (RDM) in Digital Holography for Resolution Enhancement

Caojin Yuan[1] and Hongchen Zhai[2]
[1]*Institut für Technische Optik, Universität Stuttgart, Germany*
[2]*Ministry of Education of China and Institute of Modern Optics, Nankai University, China*

5.1.1 Introduction of the Recording Plane Division Multiplexing Technique

The holographic multiplexing technique was proposed by Lloyd Cross [1] for display purposes, by which a series of holograms of a dynamic process are consequently recorded on different narrow strips of a large hologram. This multiplexing technique, namely, the SM (spatial multiplexing), together with ADM (angular division multiplexing), WDM (wavelength division multiplexing) and PM (polarization multiplexing) which are

Digital Holography for MEMS and Microsystem Metrology, First Edition. Edited by Anand Asundi.

commonly used in digital holography nowadays, are defined as the recording plane division multiplexing (RDM) in this chapter. Through those multiplexing techniques, different optical information is recorded in an efficient and optimized way.

In Section 5.1.2, the RDM technique will be implemented in pulsed digital holography, in which ADM and WDM are respectively used for ultra-fast recording with a high time resolution. RDM will be implemented in the resolution enhancement of the digital holography, in which ADM and PM are employed to record the whole information of the object, as introduced in Section 5.1.3. Conclusion and a discussion will be given in the final section.

5.1.1.1 The SM Technique

Figure 5.1.1 shows the principle of spatial multiplexing. Different optical information O_n interferes with one reference beam, producing several sub-holograms, which are superimposed on one frame of CCD. The recording process is depicted in Figure 5.1.1a, where light beams illuminate objects, respectively, with different incidental angles and the diffraction fields O_n fall onto different portions of CCD. The distributions of O_n on the recording plane should be carefully designed to avoid cross-talk in the spatial domain or in the frequency domain, as shown in Figure 5.1.1b.

Assume that each optical information, O_n, impinging upon the recording plane x–y can be expressed as:

$$O_n = O_n(x + X_n, y + Y_n) \qquad n = 1, 2, 3 \ldots \tag{5.1.1}$$

where X_n and Y_n are the offsets between the center of O_n and the coordinate origin of the recording plane, and the reference beam R is:

$$R = R_0 \exp\left[-j\frac{2\pi}{\lambda}(\sin\alpha x + \sin\beta y)\right], \tag{5.1.2}$$

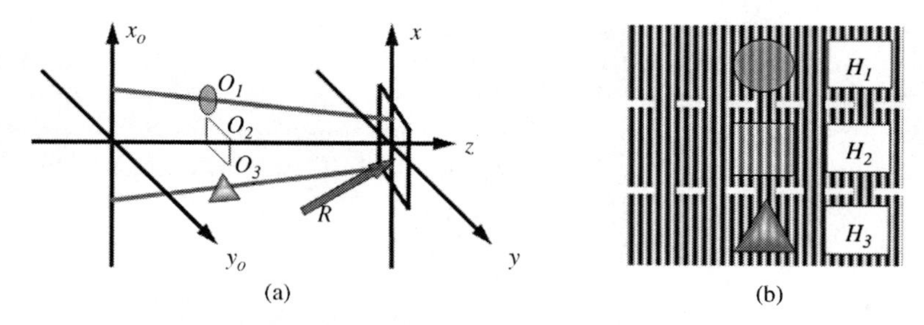

Figure 5.1.1 SM in the digital holography

where α and β are the direction angles of the reference beam with respect to the x- and y-axes respectively, then the intensity of the composite hologram will be:

$$\begin{aligned} I &= \sum_n |O_n + R|^2 = \sum_n H_n \\ &= \sum_n |O_n(x + X_n, y + Y_n) + R|^2. \end{aligned} \tag{5.1.3}$$

The recorded information O_n can be respectively reconstructed from the sub-holograms H_n by spatially dividing the composite hologram I or by Fourier transformation and digital filtering. To retrieve O_n respectively without overlapping, the dimension or the frequency bandwidth of the recorded information should be restricted to a small scale. In other words, the spatial multiplexing technique is generally used to record objects whose sizes and frequency bandwidths are limited.

5.1.1.2 The ADM Technique

Figure 5.1.2a shows the principle of multiple holographic recordings in a single frame of CCD by using ADM, where a series of incident beams with different incident spatial angles, as reference beams, are employed to interfere with a series of beams with the same incident angles containing different object

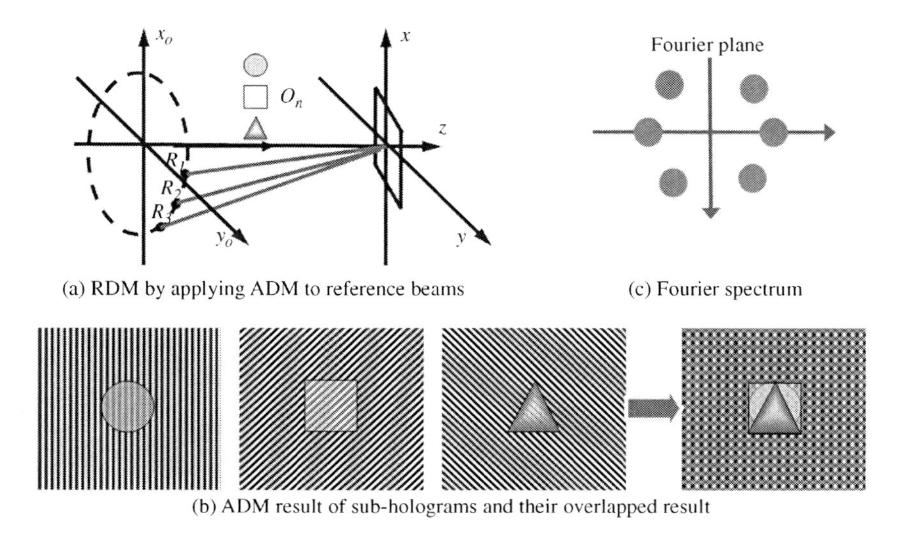

Figure 5.1.2 ADM in pulsed digital holography. (a) RDM by applying ADM to reference beams; (b) ADM result of sub-holograms and their overlapped result; (c) Fourier spectrum

information, and the sub-holograms recorded will overlap on a single frame of CCD, as shown in Figure 5.1.2b. We can use the pulsed laser source or different polarized multiplexing to avoid unwanted interference between object beams and also the interference between reference beams. In the digital reconstruction process, the holograms will be Fourier-transformed and filtered out separated spectra as shown in Figure 5.1.2c. Since the spectrum of the objects are separately in the Fourier domain, the recorded information O_n can be reconstructed and displayed respectively, through inverse Fourier transformation and diffraction calculation.

Suppose that each of the diffraction field of objects beam, O_n, on the recording plane interferes with its corresponding reference beam, R_n, with different incident angle

$$R_n = R_0 \exp\left[-j\frac{2\pi}{\lambda}(\sin\alpha_n x + \sin\beta_n y)\right] \qquad n = 1, 2, 3 \ldots \tag{5.1.4}$$

then the intensity of the composite hologram is:

$$\begin{aligned} I &= \sum_n |O_n + R_n|^2 \\ &= \sum_n \left\{ O_n^2 + R_n^2 + O_n \times R_0 \exp\left[j\frac{2\pi}{\lambda}(\sin\alpha_n x + \sin\beta_n y)\right] + \right. \\ &\quad \left. O_n^* \times R_0 \exp\left[-j\frac{2\pi}{\lambda}(\sin\alpha_n x + \sin\beta_n y)\right]\right\} \end{aligned} \tag{5.1.5}$$

The Fourier transform of the composite hologram can be expressed as:

$$\begin{aligned} FT(I) &= \sum_n FT\{O_n^2 + R_n^2 + O_n \times R_0 \exp[j2\pi(\sin\alpha_n x + \sin\beta_n y)/\lambda] + \\ &\quad O_n^* \times R_0 \exp[-j2\pi(\sin\alpha_n x + \sin\beta_n y)/\lambda]\} \\ &= \sum_n \{FT[O_n^2 + R_n^2] + \widetilde{O}_n(f_x, f_y) \otimes \delta(f_x - \sin\alpha_n/\lambda, f_y - \sin\beta_n/\lambda) + \\ &\quad \widetilde{O}_n^*(f_x, f_y) \otimes \delta(f_x + \sin\alpha_n/\lambda, f_y + \sin\beta_n/\lambda)\} \\ &= \sum_n \{FT[O_n^2 + R_n^2] + \widetilde{O}_n(f_x - \sin\alpha_n/\lambda, f_y - \sin\beta_n/\lambda) + \\ &\quad \widetilde{O}_n^*(f_x + \sin\alpha_n/\lambda, f_y + \sin\beta_n/\lambda)\} \end{aligned} \tag{5.1.6}$$

where FT is the two-dimensional Fourier transform and $\widetilde{O}_n$ is the Fourier transform of the object diffraction distribution. By choosing a suitable window function, $\widetilde{O}_n$ can be obtained after filtering, from which the individual object wave O_n can be then constructed after an inverse Fourier transform as:

$$O_n = F^{-1}\{\widetilde{O}_n(f_x, f_y)\} \tag{5.1.7}$$

Since each sub-hologram can fully utilize the recording area of CCD, ADM is an efficient multiplexing way compared to SM. However, pulsed laser, short coherence laser or other tools should be used in the recording process to avoid unwanted overlap.

5.1.1.3 The WDM Technique

Figure 5.1.3 shows the principle of RDM recordings of WDM on a single frame of a CCD, where multiple objects and reference beams with different wavelengths are employed to record multiple sub-holograms overlapped on a single frame of a CCD.

Suppose each of the object diffraction field O_n, in the recording plane can be written as:

$$O_n = O_n(x, y) \qquad n = 1, 2, 3\ldots \tag{5.1.8}$$

and the reference beam R_n is:

$$R_n = R_0 \exp\left[-j\frac{2\pi}{\lambda_n}(\sin\alpha x + \sin\beta y)\right], \tag{5.1.9}$$

where λ_n is the wavelength of nth illumination beam, the intensity of the composite hologram will be:

$$I = \sum_n |O_n + R_n|^2 = \sum_n H_n \tag{5.1.10}$$

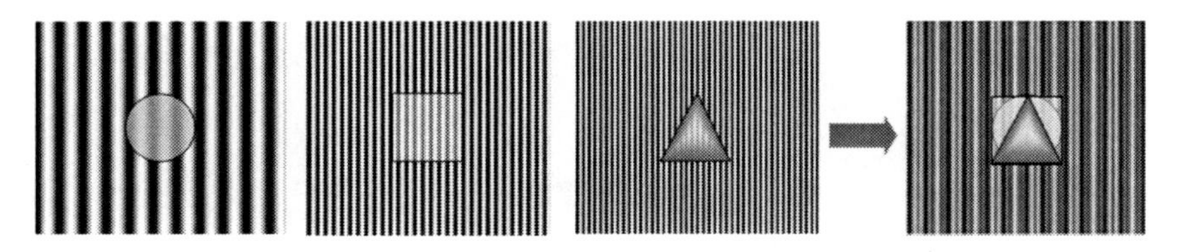

Figure 5.1.3 WDM result of sub-holograms and their overlapped result

The Fourier transform of the composite hologram can be expressed as:

$$\begin{aligned} FT(I) &= \sum_n FT\{O_n^2 + R_n^2 + O_n \times R_0 \exp[j2\pi(\sin\alpha x + \sin\beta y)/\lambda_n] + \\ &\quad O_n^* \times R_0 \exp[-j2\pi(\sin\alpha_n x + \sin\beta_n y)/\lambda_n]\} \\ &= \sum_n \{FT[O_n^2 + R_n^2] + \widetilde{O}_n(f_x, f_y) \otimes \delta(f_x - \sin\alpha/\lambda_n, f_y - \sin\beta/\lambda_n) + \\ &\quad \widetilde{O}_n^*(f_x, f_y) \otimes \delta(f_x + \sin\alpha/\lambda_n, f_y + \sin\beta/\lambda_n)\} \\ &= \sum_n \{FT[O_n^2 + R_n^2] + \widetilde{O}_n(f_x - \sin\alpha/\lambda_n, f_y - \sin\beta/\lambda_n) + \\ &\quad \widetilde{O}_n^*(f_x + \sin\alpha/\lambda_n, f_y + \sin\beta/\lambda_n)\} \end{aligned} \tag{5.1.11}$$

By the same filtering process as used in ADM, the individual object wave O_n can then be constructed after an inverse Fourier transformation as:

$$O_n = F^{-1}\{\widetilde{O}_n(f_x, f_y)\} \tag{5.1.12}$$

Each sub-hologram uses the full recording area of the CCD by using WDM. Therefore, it is also an efficient multiplexing way. However, multiple laser sources with different wavelengths or a tunable laser source are not always available.

5.1.1.4 The PM Technique

Polarization multiplexing uses the characteristic that two linearly polarized beams with orthogonal states do not interfere with each other. Therefore, two sub-holograms can be recorded simultaneously in one frame of CCD by using PM. Reference beams having different carrier frequencies and orthogonal polarization states can be expressed in terms of the Jones vectors $\vec{J}_1$ and $\vec{J}_2$

$$\vec{R}_1(x, y) = \exp\left[-i\frac{2\pi}{\lambda}(x\cos\alpha_1 + y\cos\beta_1)\right]\vec{J}_1 = R_1(x, y)\vec{J}_2 \tag{5.1.13}$$

and

$$\vec{R}_2(x, y) = \exp\left[-i\frac{2\pi}{\lambda}(x\cos\alpha_2 + y\cos\beta_2)\right]\vec{J}_2 = R_2(x, y)\vec{J}_2 \tag{5.1.14}$$

where $\vec{J}_1 = \begin{bmatrix} 0 \\ 1 \end{bmatrix}$ and $\vec{J}_2 = \begin{bmatrix} 1 \\ 0 \end{bmatrix}$.

The object diffraction distributions can be given as:

$$\vec{O}_n(x, y) = O_n(x, y)\vec{J}_n \qquad n = 1 \text{ and } 2 \tag{5.1.15}$$

The intensity distribution of the compound hologram recorded by the digital camera is presented as:

$$\begin{aligned} I &= \left|\left|\vec{R}_1 + \vec{R}_2 + \vec{O}_1 + \vec{O}_2\right|\right|^2 \\ &= \left\langle \left(\vec{R}_1 + \vec{R}_2 + \vec{O}_1 + \vec{O}_2\right), \left(\vec{R}_1 + \vec{R}_2 + \vec{O}_1 + \vec{O}_2\right) \right\rangle \end{aligned} \tag{5.1.16}$$

where $\langle \cdot, \cdot \rangle$ denotes the inner product of two vectors. Since $\left\langle \vec{J}_1, \vec{J}_2 \right\rangle = 0$, we get:

$$\begin{aligned} I &= D(x, y) + O_1(x, y)R_1^*(x, y) + O_1^*(x, y)R_1(x, y) + \\ &\quad O_2(x, y)R_2^*(x, y) + O_2^*(x, y)R_2(x, y) \end{aligned} \tag{5.1.17}$$

where $D(x, y)$ is the dc component. The O_n can reconstructed by using inverse Fourier Transform, as mentioned in Section 5.1.2.

Although it is an efficient multiplexing method, only two sub-holograms can be recorded simultaneously without unwanted interference.

5.1.2 RDM Implemented in Pulsed Digital Holography for Ultra-Fast Recording

5.1.2.1 Introduction

The pulsed digital holography technique is an effective tool for studying ultra-fast events, for it can provide amplitude and phase information of those events simultaneously. To record an ultra-fast event shorter than a CCD frame rate, multiplexing techniques can be used to record several sub-holograms in one single frame of CCD, and each sub-hologram can be reconstructed through Fourier transformation and digital filtering, as reported in [2] and [3]. In [2], a specially designed cavity is used to generate sub-pluses for object and reference beams, by which the frame interval of the recording is limited to the pico-second order. In [3], an on-axis digital holographic system with SM for recording ultra-fast event of the femtosecond order has been reported, in which the big challenge, however, is to arrange those sub-holograms in a small target of a CCD without any overlapping. The pulsed digital holography recording an ultra-fast event of the femtosecond order, with the same view angle using ADM and WDM, is reported in the following section. It is noted that the two systems make them possible to record and reconstruct the amplitude and the phase information of a dynamic process of laser-induced ionization of ambient air.

5.1.2.2 AMD in the Pulsed Digital Holography

The experimental system [4, 5] of AMD is shown in Figure 5.1.4, where the laser pulse output from a Ti: sapphire laser amplifier system combined with a half-wave plate is divided by a polarizing beam splitter into two beams, namely, the exciting beam and the recording beam. The former is focused by lens L to excite an ultra-fast event, and the latter is further divided by the beam splitter BS_1 into two parts, for SPG_2 to generate that of reference beams and for SPG_1 to generate the sub-pulse-train of the object beams, respectively. Beam splitter BS_2 is used to couple the object beams and the reference beams into the recording plane of the CCD. There is a set of mirrors and beam splitters in each SPG, which can adjust the sub-pulse-train of the object and reference beams with the same time delays, but the orientation of each can be respectively adjusted. All the incident sub-pulses upon the recording plane from SPG_1 will have different spatial angles, while the beams upon the event to be recorded as well as that upon the recording plane from SPG_2 will be kept the same, as shown in Figure 5.1.4. This can ensure a successful RDM recording object at the same viewing angle. Furthermore, by tuning the mechanical stage over a range of μm to mm, the time delay between the sub-pulses in a pulse train can be adjusted from 300fs to the ps order. The 4f system composed of L_1 and L_2 is used to record amplified image holograms at the recording plane of a single frame of the CCD.

In the experiment, the focus length of lenses L_1 and L_2 is 15 mm and 150 mm, respectively, with which a 10 times magnified image hologram of the ionization region can be recorded by a CCD with a pixel number of 576×768 and a pixel size of $10.8\,\mu m \times 10\,\mu m$, respectively. A Ti: sapphire laser amplifier system (Spitfire HP 50), with the maximum output energy of single pulse of 2 mJ, a repeated frequncy of 1 kHz, and a FWHM (Full Width at Half Maximum,) of 50 fs at a wavelength of 800 nm, is used as an optical source for inducing both the ultra-fast event of air ionization and its holographic recording. The

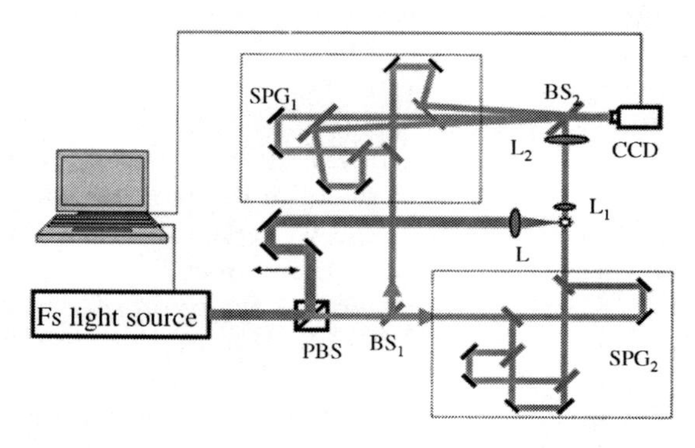

Figure 5.1.4 The recording system with ADM

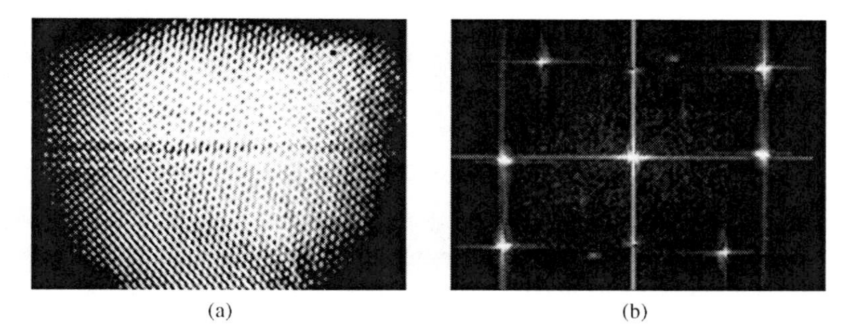

Figure 5.1.5 Composite hologram (a) and its Fourier frequency spectra (b) (Reproduced by permission of the Society of Photo-optical Instrumentation Engineers © 2007)

amplifier system and the CCD image capturer are synchronously controlled by a computer, to have the output of a single laser pulse match the recording of a single frame of CCD in the time scale.

The composite hologram and its Fourier frequency distribution are shown in Figures 5.1.5a and b, respectively. It is noted that, by employing ADM, three Fourier frequency spectra of the sub-holograms along different orientations are spatially separated in the Fourier plane. Their reconstructed intensity and phase images are shown in Figure 5.1.6, where the exposure

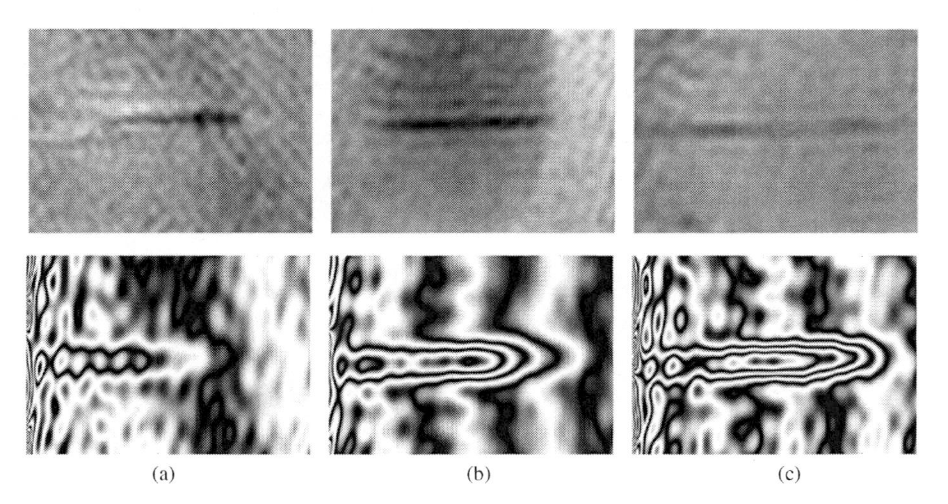

Figure 5.1.6 Intensity and phase images digitally reconstructed from the sub-hologram shown in Figure 5.1.5 with an exposure time of 50 fs and the frame interval $\Delta t_1 = 300$ fs between frames (a) and (b) and $\Delta t_2 = 550$ fs between frames (b) and (c), respectively (Reproduced by permission of the Society of Photo-optical Instrumentation Engineers © 2007)

time is 50 fs and the frame intervals are adjusted from 300 to 550 fs, demonstrating clearly the gradient increase of electron density within the plasma of the femtosecond order.

5.1.2.3 WDM in Pulsed Digital Holography

WDM can be also employed in pulsed digital holography to record ultra-fast events, while keeping the incident angles the same. The sub-holograms with different wavelengths can be recorded in the same frame of a CCD with a different spatial frequency, which can easily be separated through spatial frequency filtering in the Fourier domain during the digital reconstruction process, in the same way as in the case of ADM, under an appropriate experiment design.

The experiment system [6] is shown in Figure 5.1.7, where the laser pulse from a Ti: sapphire laser amplifier system is divided by a polarizing beam splitter PBS with a half wave plate into two parts: the exciting beam and the probing beam. The former can be focused by lens L to excite air ionization, and the latter passes through the halfwave plate P1 and BBO crystal to generate harmonic wave. The half wave plate is used to adjust the polarization of the incident beam to obtain frequency doubling efficiency. The basic and frequency doubled waves will be separated by a dichroic mirror DM1 into two parts with a time delay, so that two sub-holograms based on WDM can be recorded at different times. From the dichroic mirror DM2, the successive optical path based on Michelson's interferometer is equal for both of the two wavelengths, in which M3 and M4 are used to ensure the same optical path in the

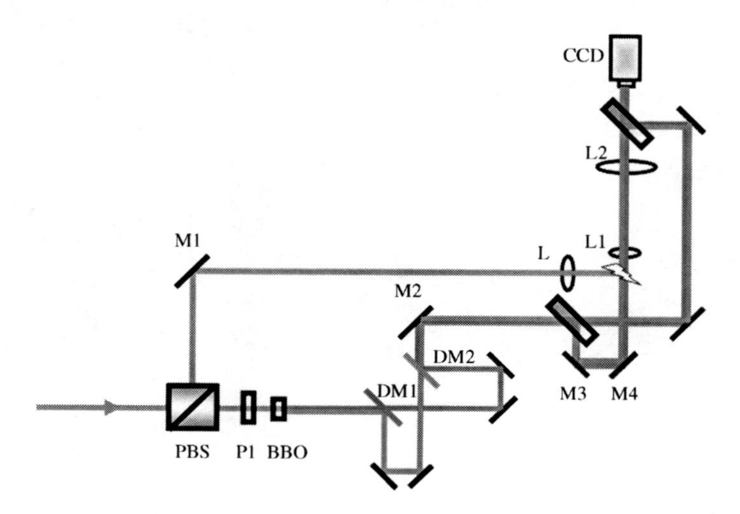

Figure 5.1.7 The experimental set-up with WDM

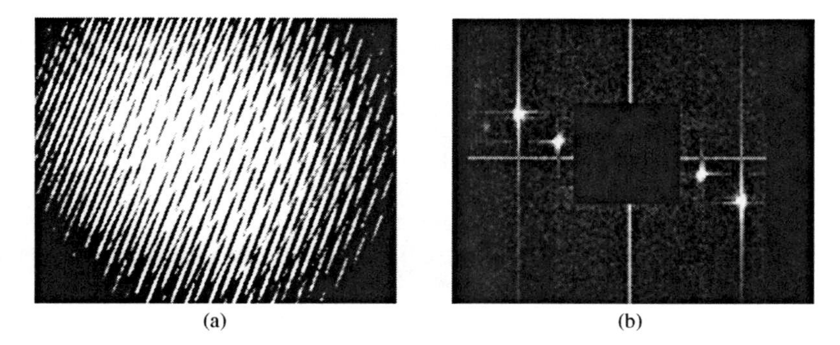

Figure 5.1.8 Overlapped sub-holograms (a) and their Fourier frequency spectra (b) (Reproduced by permission of the Society of Photo-optical Instrumentation Engineers © 2007)

object beam and reference beam arms. The 4f system composed of L1 and L2 is used to record both amplitude and phase images on the recording plane of the CCD. The parameters of the laser source and digital camera are the same as the above experiment.

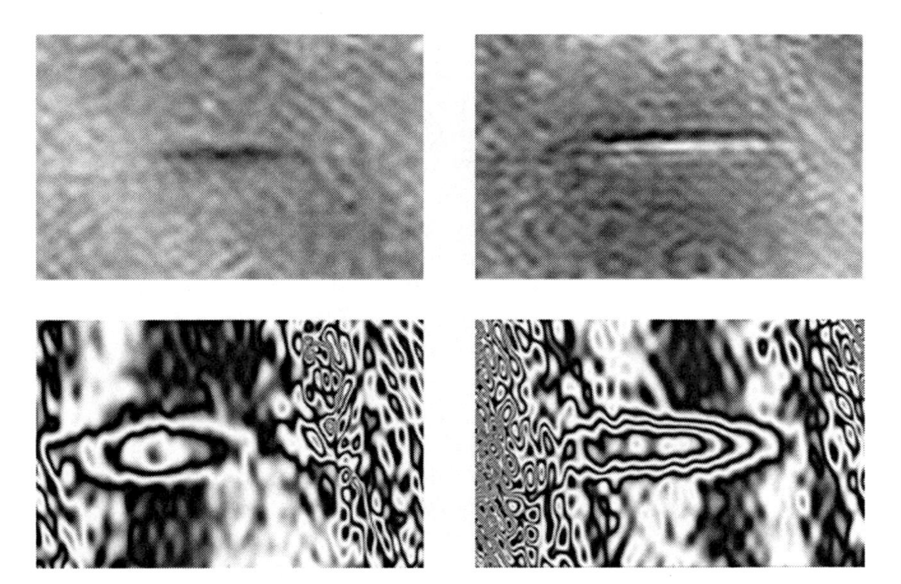

Figure 5.1.9 Reconstructed intensity and phase images of air ionization (Reproduced by permission of the Society of Photo-optical Instrumentation Engineers © 2007)
Note: The energy and width of pump laser pulse are 0.4 mJ and 50 fs, respectively. The frame interval is about 400 fs

To demonstrate the possibility of recording the ultra-fast process of air ionization by WDM, the composite hologram containing two sub-holograms as well as its Fourier spectrum are shown in Figures 5.1.8a and b, respectively. It is noticed that, by employing WDM, two Fourier spectra of the sub-holograms are spatially well separated. After filtering the sub-holograms in the Fourier spectrum plane, followed by an inverse Fourier transform, the complex amplitude of the reconstructed images can be obtained. Their reconstructed intensity and phase images are shown in Figure 5.1.9, where the exposure time is 50 fs and the time delay between exposures is 400 fs, which clearly shows an area of the same electron density within the plasma induced by the femtosecond laser pulse.

5.1.3 RDM Implemented by Digital Holography for Spatial Resolution Enhancement

5.1.3.1 Introduction

Digital holography is a promising measurement method, which is widely used in microscopy, deformation measurement, three-dimensional reconstruction, and so on [7, 8]. The resolution of a digital holographic recording system is related to the Numerical Aperture (NA) and the wavelength (λ) of the light source. One of the effective ways to enlarge NA with a given wavelength and finally to improve the resolution, is to record a hologram with a larger dimension CCD or at a short distance to the recorded object, which is limited by the available device and the digital sampling requirement, respectively. Introducing the microscope objective into the digital holographic system is another way to achieve a high resolution, of which the resolution is equal to that of its microscopic imaging system and is given by $0.77\lambda/NA$. High resolution can be achieved by choosing a microscope objective with a large NA. However, a large NA microscope objective is associated with a small field of view, a short working distance and focus depth. The aperture-synthesis approach is an alternative solution to improve the resolution, for which a set of sub-holograms are recorded at different positions [9]. However, it is time-consuming and requires a lot of system stability. Modified synthetic aperture digital holography has been studied to obtain phase and amplitude contrast super-resolution images, where the low frequency information of the object obtained by on-axis illumination and the high frequency information produced by off-axis illuminations can be simultaneously recorded [10, 11]. To avoid the interference between low and high frequency information during the recording process, one needs to employ other techniques or tools such as vertical-cavity surface-emitting lasers (VCSEL), filter masks or orthogonal polarization states. DH systems combining SM with filtering masks are simple, but the parameters of filtering masks are difficult to adjust according to different objects.

In this section, we describe two resolution enhancement methods with RDM in digital holography, where ADM with ultra-short pulsed laser source, and ADM and PM are respectively used in DH. From the experimental results, it is shown that the resolutions of the reconstructed images are improved and exceed the resolutions determined by the numerical aperture of the recording systems.

5.1.3.2 AMD in Digital Holography

A synthetic aperture holographic recording system [12] is shown in Figure 5.1.10, where a single laser pulse output from a pulsed laser system is divided into two parts, of which one is divided into three sub-pulses with an equal time delay. The object beams with different incident angles, as on-axis and off-axis illuminations, make three diffraction object fields record the CCD, respectively, which contain the low and high frequency information of the object. The other part will also be further divided into three sub-

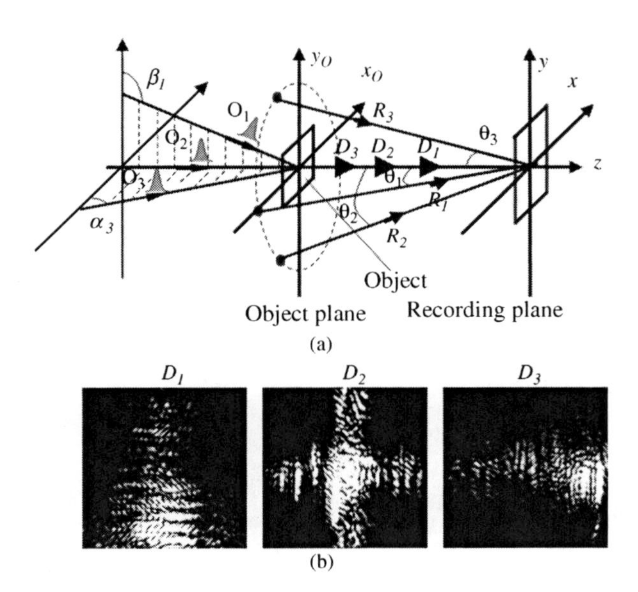

Figure 5.1.10 Object illumination O_n, its diffraction D_n and reference beams R_n ($n = 1, 2, 3$). (a) Sketch of the relation of O_n, D_n and R_n. (b) Different distribution of D_n in recording plane (Reproduced by permission of © 2008 Optical Society of America)

pulses in the same way as reference beams, so that three pairs of incidence beams can be recorded as three sub-holograms with different orientations of the interference fringes.

As shown in Figure 5.1.10a, the object and the CCD are located in $x_O - y_O$ plane and x-y plane, respectively. Under the Fresnel approximation, if the distance between the object plane and the recording plane Z_O is large enough, the diffraction field distribution of the object on (x, y) plane will be:

$$O_n(x,y) = C'\text{rect}\left(\frac{x}{L_x}\right)\text{rect}\left(\frac{y}{L_y}\right)\frac{\exp(ikZ_O)}{i\lambda Z_O}\exp\left[\frac{i\pi}{\lambda Z_O}(x^2+y^2)\right]\iint t(x_O,y_O)\times \exp\left[\frac{i\pi}{\lambda Z_O}(x_O^2+y_O^2)\right]\exp\left\{-i2\pi\left[x_O\left(\frac{x}{\lambda Z_O}-\frac{\cos\alpha_n}{\lambda}\right) +y_O\left(\frac{y}{\lambda Z_O}-\frac{\cos\beta_n}{\lambda}\right)\right]\right\}dx_O dy_O \tag{5.1.18}$$

where the rectangle function is defined by the dimension of the CCD L_x and L_y, α_n and β_n is the angle included between the illumination direction vector and x_O- and y_O-axes, and $t(x_O, y_O)$ is the transmittance of the object. Equation 5.1.18 shows that the diffraction field distribution of the object $O_n(x, y)$ can be regarded as the angular spectrum of the object transmittance $t(x_O, y_O)$ multiplied by a quadratic phase factor with a coordinate shift of the value of $(\cos\alpha_n/\lambda, \cos\beta_n/\lambda)$.

Under on-axis illumination, as mentioned above, the highest spatial frequency that can be recorded is limited by NA/λ. Since three sub-holograms recorded on the same frame of CCD cover three different spatial frequency ranges of the object, the synthesized resolution will be higher than that of recording only one hologram under on-axis illumination. After synthesizing the low and high frequency information, the maximum spectral range in the x and y direction is $\left[-\frac{NA}{\lambda},\frac{NA}{\lambda}+\frac{\cos\alpha_n}{\lambda}\right]$ and $\left[-\frac{NA}{\lambda},\frac{NA}{\lambda}+\frac{\cos\beta_n}{\lambda}\right]$, respectively. Meanwhile, either the higher or the lower spatial frequency portion of the object diffraction will fall on the CCD recording area by choosing a proper off-axis or on-axis object illumination, respectively.

In the recording process, the relative spatial angle included between the object and each reference beam, θ_1 to θ_3 in Figure 5.1.10a, is chosen as $\theta_1 = 1.181°$, $\theta_2 = 1.161°$ and $\theta_3 = 1.006°$, respectively, the recording distance is adjusted as $Z_O = 52$ mm, and the angle of the y-off-axis, on-axis, and x-off-axis object illumination O_n is selected as ($\alpha_1 = 90°$, $\beta_1 = 92.9°$), ($\alpha_2 = 90°$, $\beta_2 = 90°$), and ($\alpha_3 = 85.9°$, $\beta_3 = 90°$), respectively, to ensure that the high and the low spatial frequency components of the object diffraction can be respectively recorded in one of the three sub-holograms.

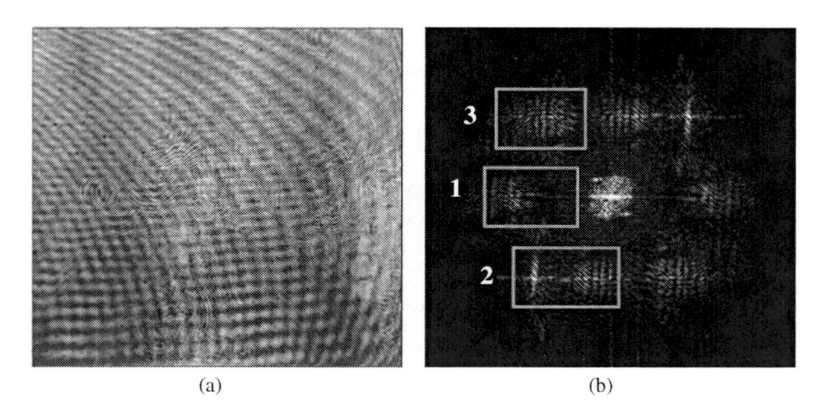

Figure 5.1.11 The composite hologram (b) and its frequency spectrum (b) (Reproduced by permission of © 2008 Optical Society of America)

The three overlapping sub-holograms and their Fourier spectra are shown in Figures 5.1.11a and b, respectively. It is noticed that the Fourier transform spectra of the three sub-holograms are separated, in which the separated +1 spectra orders are marked by rectangles 1 to 3 in Figure 5.1.11b. After the frequency filtering followed by an inverse Fourier transform of +1 spectra order of each sub-hologram, three object diffraction fields $O_n(x, y)$ can be obtained, and the object wavefront in the x_O-y_O plane can be obtained by calculating the Fresnel diffraction and removing the carrier phases introduced by the object illumination beams. The reconstructed transmittances of the object and their magnified parts with higher resolution are shown in Figure 5.1.12. The measured maximum resolution in the

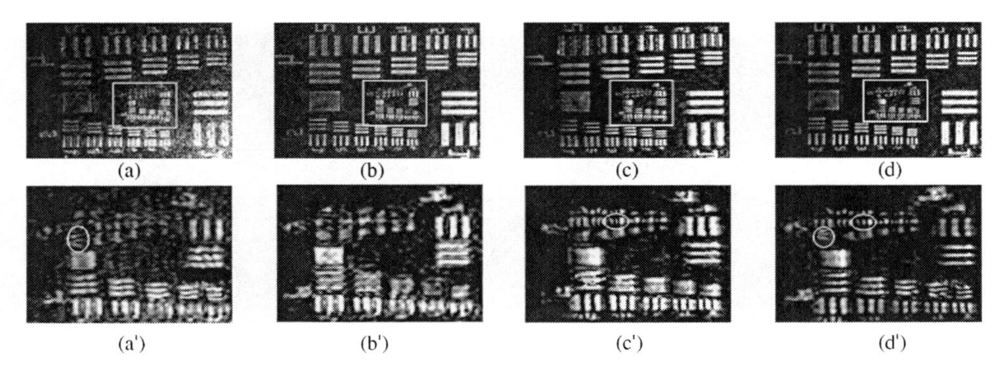

Figure 5.1.12 Intensity distribution of the reconstructed images of O_n (n = 1,2,3) object illumination respectively (a–c), the aperture-synthesized image (d), and their central parts enclosed in the rectangles (a′–d′) (Reproduced by permission of © 2008 Optical Society of America)

vertical and horizontal direction of the reconstructed test target can be found in the circled part of Figure 5.1.12a′, 128.00 Lp/mm obtained by y-off-axis object illumination and in that of Figure 5.1.12c′, 161.30 Lp/mm obtained by x-off-axis object illumination, respectively.

The intensity image of the object obtained by synthesizing the three complex amplitudes is shown in Figure 5.1.12d and its magnified image is given in Figure 5.1.12d′. By comparing Figure 5.1.12d′ with the conventionally reconstructed image of on-axis illumination in Figure 5.1.12b′, it is found that, while keeping the quality of the low frequency section of the reconstructed image unchanged, the maximum resolution has been improved from 71.84 Lp/mm to 128.00 Lp/mm in the vertical direction and 90.51 Lp/mm to 161.30 Lp/mm in the horizontal direction. Meanwhile, those experimental data are close to the theoretical values which are improved from 74.77 Lp/mm to 138.01 Lp/mm in the vertical direction and from 92.31 Lp/mm to 181.68 Lp/mm in the horizontal direction. The difference between the experimental results and the corresponding theoretical values may be due to the discontinuous change of resolution in the test pattern and other factors like noise.

5.1.3.3 AMD and PM in Digital Holography

Angular multiplexing techniques with ultra-short pulsed laser can be used to record the high and low spatial frequency information sequentially, as mentioned above; however, the ultra-short pulsed laser is expensive and not always available. In this section we present another approach to improve the resolution combining AMD with PM, where orthogonal polarization states are used to avoid the cross-talk between the low and high frequency information, and the angular multiplexing technique is used to separate those recorded information.

The proposed microscopic digital holographic recording system [13] is shown in Figure 5.1.13. A collimated and expanded plane wave is divided into

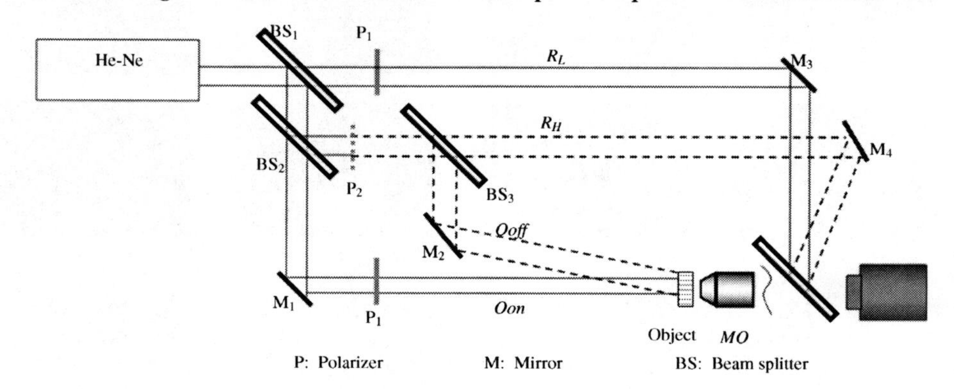

Figure 5.1.13 Experimental set-up

two beams by the beam splitter BS_1. One of those two beams serves as reference R_L, where the subscript L stands for low frequency, and the other beam is further divided into an on-axis illumination object beam O_{on} and a new beam by BS_2. We insert two identical polarizers P_1 into the reference beam R_L and on-axis illumination, respectively, which set the two beams to be linearly polarized in the y direction. The beam reflected by BS_2 is linearly polarized in the x direction after passing through P_2. This beam is further divided by BS_3 into an off-axis illumination beam O_{off} and another reference beam R_H, where the subscript H stands for high frequency. The linear polarization states in the x and y direction are denoted as dashed and solid lines in Figure 5.1.13, respectively. The object is illuminated by the beams O_{on} and O_{off} with two different incident angles and imaged on the camera. For the case of nonbirefringent samples, the two incident beams O_{on} and O_{off} transmitting through the object, can only interfere with their corresponding reference beams (R_L and R_H) respectively, because of their mutual orthogonal polarization states. Therefore, only two interference patterns (two sub-holograms) are formed.

During the experiment, the light source used is a He-Ne laser with the wavelength $\lambda = 632.8\,\text{nm}$. The compound hologram is recorded by a CMOS with 2048(H) $\times$ 1536(V) pixels; the pixel size is $6.554\,\mu\text{m} \times 4.915\,\mu\text{m}$. The direction angle of the off-axis illumination beam to the x_O- and the y_O-axis is $\alpha_{oH} = 87^\circ$ and $\beta_{oH} = 90^\circ$, respectively. A 5$\times$ microscope objective with NA = 0.15 is used to image the object onto the digital camera. With these parameters, the resolution of the system is $0.77\lambda/\text{NA} = 3.25\,\mu\text{m}$, which corresponds to the No. 2 group in 7th element of the USAF1951 test target.

A compound hologram recorded with the set-up is shown in Figure 5.1.14a, where a portion included in the solid lines is magnified to show the interference fringes. The Fourier transformation of the hologram is shown in Figure 5.1.14b. From this figure, it can be seen that the O_L and O_H as well as

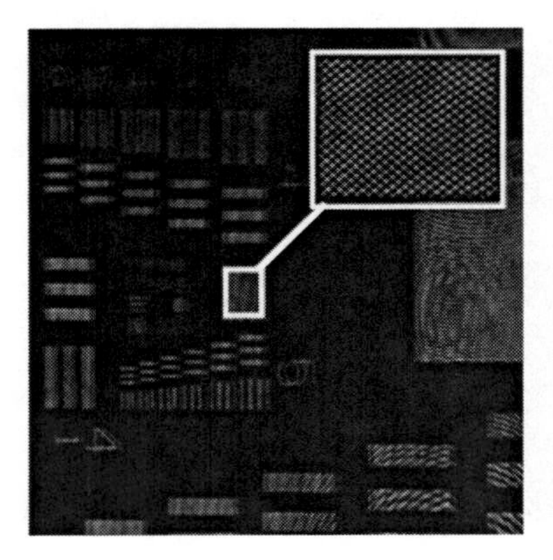

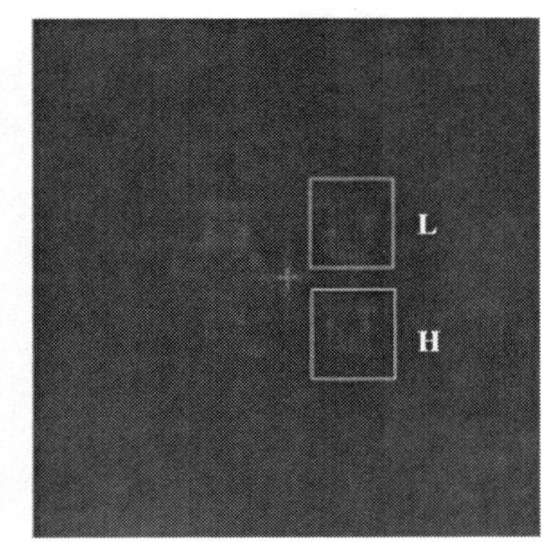

Figure 5.1.14 Compound hologram (a) and its spectrum (b) (Reproduced by permission of © 2011 Optical Society of America)

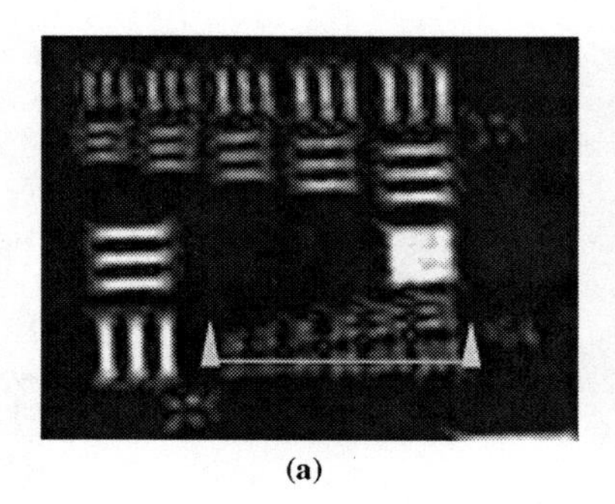
(a)

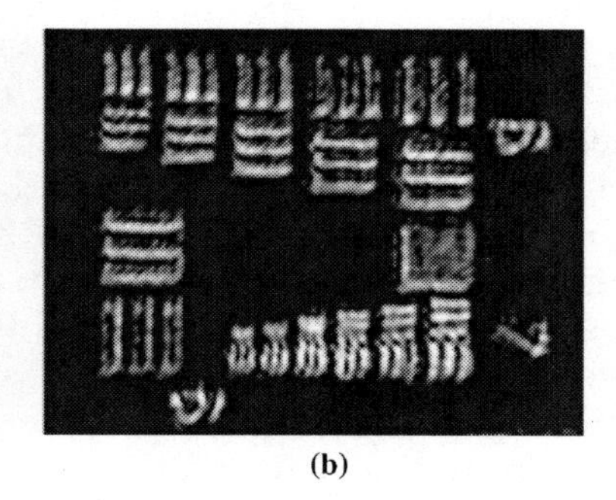
(b)

Figure 5.1.15 Intensity distribution of the reconstructed images containing low (a) and high (b) frequencies (Reproduced by permission of © 2011 Optical Society of America)

their conjugated wavefronts are separated. After frequency filtering and inverse Fourier transform, the reconstructed images covering the low and high frequency labeled as "L" and "H" in Figure 5.1.14b are shown in Figures 5.1.15a and b, respectively. Although under off-axis illumination the patterns of the 7th group in the vertical direction can be resolved, the patterns in the 6th group are of poor quality because of the lack of low frequencies information. On the other hand, the patterns in the 6th group have uniform intensity, but the patterns in the 7th group cannot be resolved as shown in Figure 5.1.15a, due to lack of high frequencies information.

The synthesized image of the object obtained by adding the two complex amplitudes is shown in Figure 5.1.16a and its magnified image is given in Figure 5.1.16b. In order to compare the synthesized image with the reconstructed image obtained under on-axis illumination, the intensity plots along

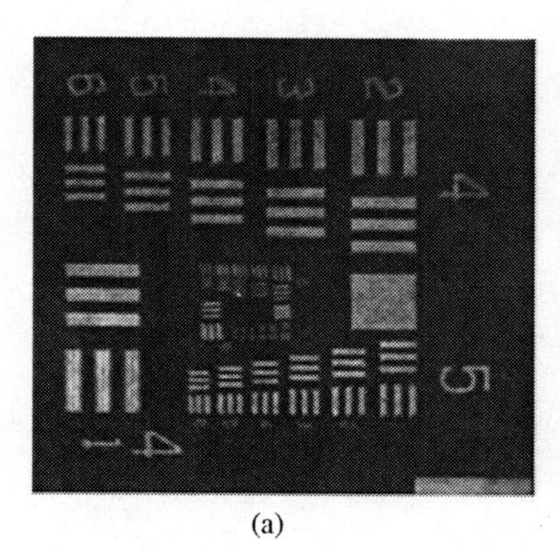
(a)

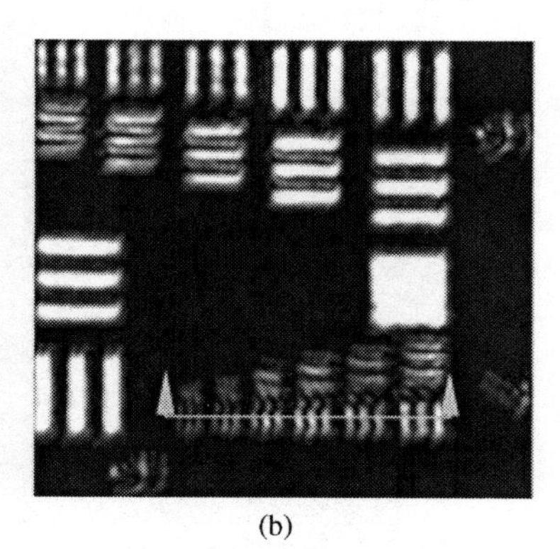
(b)

Figure 5.1.16 Intensity distribution of the synthesized image (a) and its partly magnified image (b) (Reproduced by permission of © 2011 Optical Society of America)

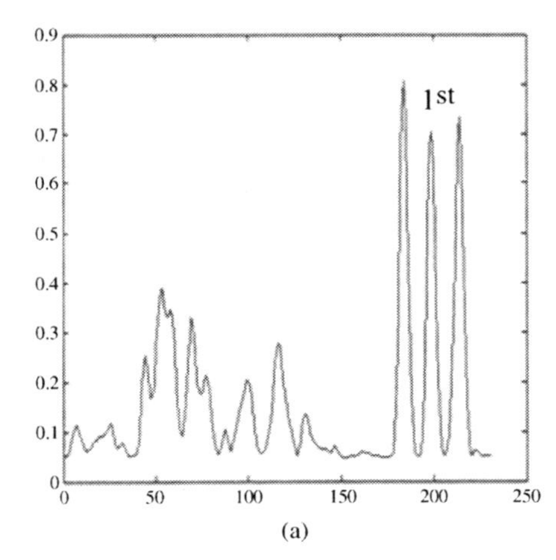

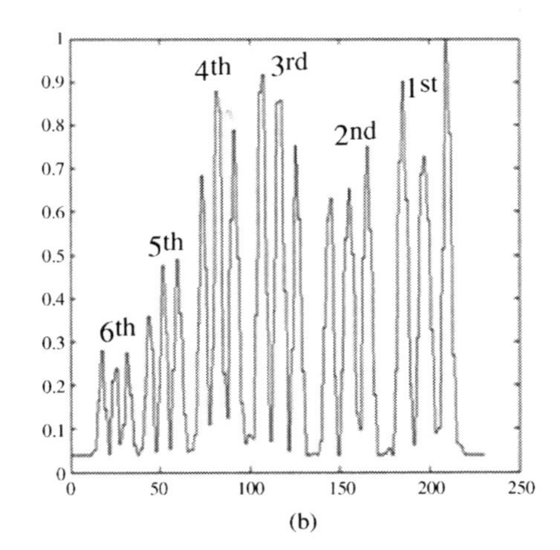

Figure 5.1.17 The plots along the white solid lines of Figures 5.1.15a and 5.1.16b (Reproduced by permission of © 2011 Optical Society of America)

the solid white line in Figures 5.1.15a and 5.1.16b are shown in Figure 5.1.17. It is clear that the maximum resolution of the high frequency portion has been improved from 128.0 Lp/mm to 228.0 Lp/mm in the vertical direction.

5.1.4 Conclusion

RDM is applied in digital holography for ultra-fast events recording and microscopic imaging, with which a series of the images in an ultra-fast process and multiple images covering different frequency bandwidths can be recorded on one frame of the digital camera. In all of the reconstruction processes of the RDM recording, Fourier transformation and frequency filtering in the Fourier plane are employed to separate the spatial spectra of the multiple recordings.

In the case of holographic recording of ultra-fast events, the optimized designs of the optical paths of the object and the reference beams in digital holographic systems with ADM or WDM make it possible for them to record ultra-fast events of the femtosecond order, while keeping the incident angle of the object beams unchanged. The amplitude and phase images with the same view angle digitally reconstructed demonstrated the ultra-fast dynamic process of laser-induced ionization of ambient air at a wavelength of 800 nm, with a time resolution of 50 fs and a femtosecond frame interval.

In the case of spatial resolution enhancement, the digital holographic system using ADM with the ultrashort laser or ADM and PM is used to record the low and high frequency information in one frame without unwanted interference, respectively. After the filtering and reconstruction algorithm, a super-resolution image can be obtained by synthesizing the complex amplitudes of the reconstructed images. Compared with the super-resolution recording systems mentioned in the introduction to Section 5.1.3, there are no filtering masks and they are more robust. The experimental results show that the resolutions of the reconstructed images with aperture synthesis are improved by a factor of 1.8 compared with that with on-axis illumination.

In the digital holographic RDM recording for the ultra-fast process, the time resolution of the recording system is limited by the laser pulse, but the frame interval in recording can be tuned by adjusting the optical path of the object and the reference beams. In the digital holographic RDM recording for spatial resolution enhancement, the resolution could be improved further if more off-axes object illumination could be adopted.

References

1. Goodman, J.W. (2005) *Introduction to Fourier Optics*, 3rd edn , Roberts & Company, Colorado.
2. Liu, Z., Centurion, M., Panotopoulos, G. *et al.* (2002) Holographic recording of fast events on a CCD camera. *Opt. Lett.*, **27** (1), 22–24.
3. Centurion, M., Pu, Y., Liu, Z. *et al.* (2004) Holographic recording of laser-induced plasma. *Opt. Lett.*, **29** (7), 772–774.
4. Wang, X., Zhai, H., and Mu, G. (2006) Pulsed digital holography system recording ultrafast process of the femtosecond order. *Opt. Lett.*, **31** (11), 1636–1638.
5. Wang, X., Zhai, H., and Mu, G. (2007) Ultra-fast digital holography of the femtosecond order. *Proc. SPIE*, **6279**, 62791E-1–62791E-6.
6. Wang, X.and Zhai, H. (2007) Pulsed digital micro-holography of femto-second order by wavelength division multiplexing. *Opt. Commun.*, **275**, 42–45.
7. Cuche, E., Marqret, P., and Depeursingle, C. (1999) Simultaneous amplitude-contrast and quantitative phase-contrast microscopy by numerical reconstruction of Fresnel off-axis holograms. *Appl. Opt.*, **38** (34), 6994–7001.
8. Mann, C., Yu, L., Lo, C., and Kim, M. (2005) High-resolution quantitative phase-contrast microscopy by digital holography. *Opt. Express*, **13** (22), 8693–8698.
9. Clerc, F.L., Gross, M., and Collot, L. (2001) Synthetic-aperture experiment in the visible with on-axis digital heterodyne holography. *Opt. Lett.*, **26** (20), 1550–1552.
10. Mico, V., Zalevsky, Z., and García, J. (2007) Synthetic aperture microscopy using off-axis illumination and polarization coding. *Opt. Commun.*, **276**, 209–217.
11. Mico, V., Zalevsky, Z., García-Martínez, P., and García, J. (2006) Superresolved imaging in digital holography by superposition of tilted wavefronts. *Appl. Opt.*, **45** (5), 822–828.
12. Yuan, C., Zhai, H., and Liu, H. (2008) Angular multiplexing in pulsed digital holography for aperture synthesis. *Opt. Lett.*, **33** (20), 2356–2358.
13. Yuan, C., Situ, G., Pedrini, G. *et al.* (2011) Resolution improvement in digital holography by angular and polarization multiplexing. *Appl. Opt.*, **50** (7), B6–B11.

5.2

Development of Digital Holographic Tomography

Yu Yingjie
Department of Precision Mechanical Engineering, Shanghai University, China

5.2.1 Introduction

Optical tomography can discover the section data effectively and measure the 3D structure of objects. Digital holographic tomography (DHT) is also an optical tomography technology, which combines digital holography with tomography. The basic thought is to apply digital holographic technology [1, 1–3] to reconstruct the object's 3D profile, for example, the phase information, and rebuild the object's 3D structure with computer tomography (CT) or optical coherence tomography (OCT). DHT has several distinct advantages, such as non-contact, a simple and stable collecting system, and a good dynamic performance. It has broad potential for the detection of heterogeneous materials with multilayer refraction index, such as functions gradient materials.

In 1969, Wolf [4] first used holographic tomography to restructure translucent objects. Subsequently, research was performed on hologram recording, 3D reconstruction algorithms and applications. Classical holographic tomography mainly measures dynamic physical quantities, such as the 3D inspection of pneumatic flow, temperature distribution, density distribution [5, 6], and

Digital Holography for MEMS and Microsystem Metrology, First Edition. Edited by Anand Asundi.

so on. In the past few years, related research has gradually been done on digital holographic tomography, such as biological tissue ingredient inspection, 3D reconstruction of the refraction index, or measurement of some physical quantities or living cell imaging [7, 7–16], and analysis of reconstruction errors [17].

5.2.2 Classification of Digital Holographic Tomography

According to different combined techniques, the hologram recording system of digital holographic tomography can be divided into two types: the holographic scanning method and the coherent slicing method. The former primarily aims at the registration of transparent material. It adopts the Mach–Zehnder holographic interferential system combined with 3D reconstruction algorithms used in CT. When recording holograms, the measured sample needs be rotated from 0 degrees to 180 degrees [8, 9]. Figure 5.2.1 shows the experimental system of DHT [8] on the basis of digital holographic microscopy conducted by E. Cuche and his research group.

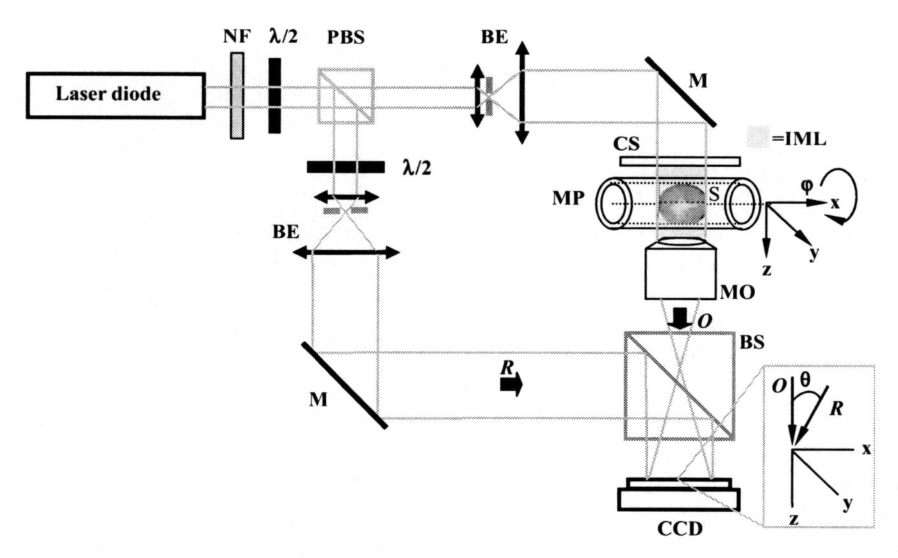

Figure 5.2.1 Holographic microscope for transmission imaging (Reproduced by permission of © 2006 Optical Society of America)
Note: NF signifies neutral-density filter; PBS signifies polarizing beam splitter; BE signifies beam expander with spatial filter; λ/2 represents half-waveplate; MO signifies microscope objective; M signifies mirror; BS signifies beam splitter; O signifies object wave; R signifies reference wave; MP signifies micropipette; S signifies specimen; IML signifies index matching liquid; Inset shows represented detail showing the off-axis geometry at the incidence on the CCD

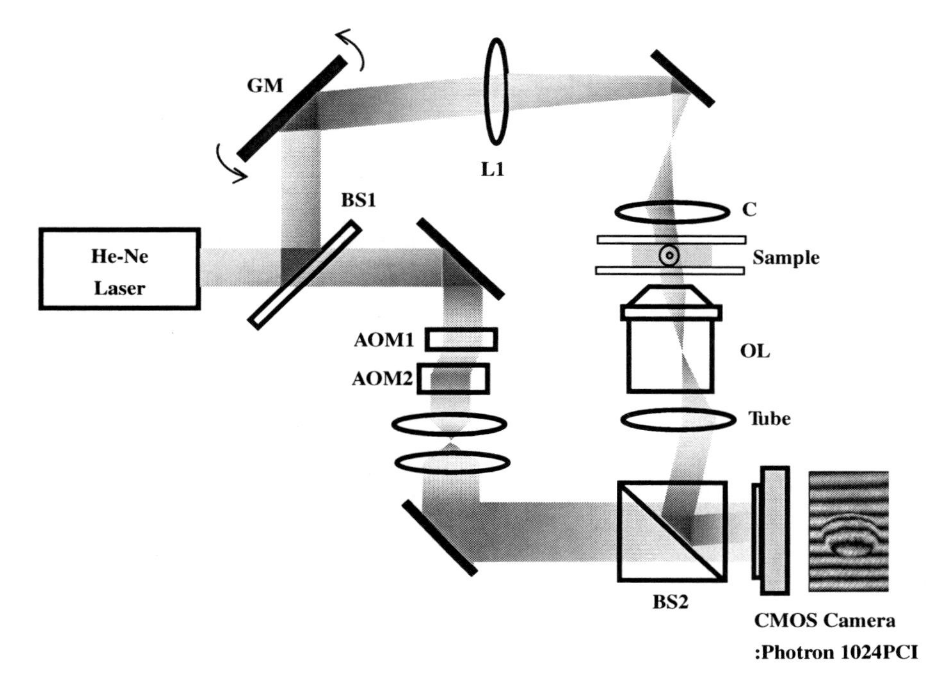

Figure 5.2.2 Tomographic phase microscope via different illumination angle (Reproduced by permission of © 2007 Nature Publishing Group)
Note: Gm signifies galvanometer scanning mirror; L1 signifies lens (f = 250 mm); C signifies condenser lens (NA 1.4); OL signifies objective lens (NA 1.4); Tube signifies tube lens (f = 200 mm); BS1 and BS2 are beam splitters; AOMs signifies acousto-optic modulators. The frequency shifted reference laser beam is shown in gray

During the experiment, the biological sample is fixed in a transparent container and rotated, a hologram is recorded every 2 degrees, and 90 holograms are collected. The 90 holograms are numerically reconstructed respectively. Finally, the measured biologic sample's 3D refraction index distribution is acquired by means of a filtered back-projection algorithm used in CT. Some researchers also suggest that the object's data can be achieved by rotating the angle of the illuminating light, as shown in Figure 5.2.2 [16].

The coherent slicing method is mainly used to record some semitransparent or reflective samples. It is based on OCT theory, combined with the Michelson interferential system and low coherent light source. Digital holograms are acquired layer by layer by the confocal scanning method. The sketch of the optical system used to achieve OCT with a short-coherence digital holographic microscope is shown in Figure 5.2.3 [7]. Multi-wavelength scanning [12,15] is also adapted to record data on different depths and to acquire the measured

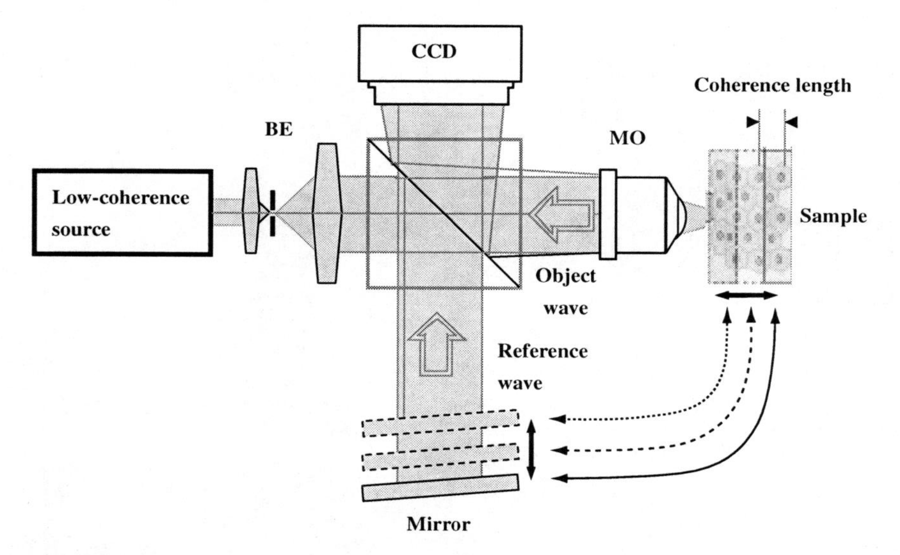

Figure 5.2.3 Sketch of the optical system used to achieve OCT with a short-coherence digital holographic microscope (Reproduced by permission of © 2005 Optical Society of America)
Note: B.E signifies beam expander; M.O. signifies microscope objective

sample's fault intensity information. Sub-micrometer optical tomography by multiple-wavelength digital holographic microscopy is shown in Figure 5.2.4.

Recently, A. Jozwicka [13, 14] has proposed a new system for scanning holographic tomography with digital off-axis Mach–Zehnder interferometry holography, as shown in Figure 5.2.5.

A collimated laser beam is passed through the transparent object from three different directions successively. The object wavefronts corresponding to the three propagation directions were recorded in a single hologram with a CCD. At the corresponding distances, same as the recording distances, data on the three directions was reconstructed. Finally, from the reconstructed information, the refractive index distributions inside the object were calculated with an algebraic iteration algorithm.

The 3D reconstruction algorithms of DHT can also be divided into two types: the transform algorithm (the convolution algorithm) and the series expansion algorithm (the algebraic reconstruction algorithm). The transform algorithm is suitable for the 3D reconstruction with complete projecting data, namely, it needs to scan the object in range of 180 degrees. The algebraic reconstruction algorithm is suitable for the 3D reconstruction with lack of sample data. The method adopted is to choose a group of basic functions and then use their suitable linear combination to approach the object function.

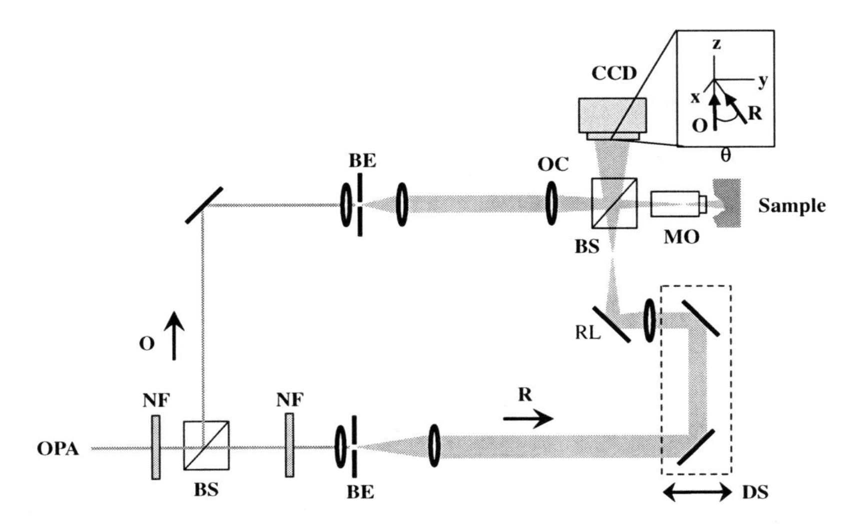

Figure 5.2.4 Sub-micrometer optical tomography by multiple-wavelength digital holographic microscopy (Reproduced by permission of © 2006 Optical Society of America)

Note: O signifies object arm; R, reference arm; OPA signifies adjustable wavelength laser; NF signifies neutral filter; BS signifies beam splitters; BE signifies beam expander; MO signifies microscope objective; OC signifies object beam condenser; RL signifies reference lens; CCD signifies charged-coupled device camera; DS signifies delay system

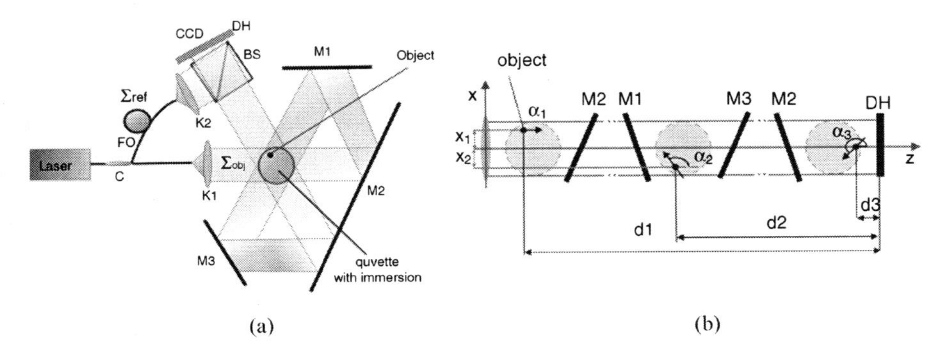

Figure 5.2.5 The concept of DHT. (a) Set-up for recording a hologram; (b) Linear unfolding of the object beam's multiple passage through the object (Reproduced by permission of Society of Photo-optical Instrumentation Engineers © 2007)

Note: FO signifies fiber optics; C signifies fiber optics coupler; K1, K2 are collimators; M1, M2, M3 are mirrors; DH signifies digital hologram's plane

5.2.3 Principle of Digital Holographic Tomography

DHT is the integration of digital holography and CT reconstruction techniques. At first, it used a digital holographic recording system to register some digital holograms, and then reconstructed the holograms to get the phase maps, which present the projecting data of the wavefronts passed through the object. The change of internal refractive index creates the variety of phase map of wavefronts, therefore these phase maps from different projection views are used as the projecting data for the CT reconstruction. The internal structure or parameters associated with multi-layer refractive indexes can be obtained by suitable CT reconstruction algorithms.

5.2.3.1 Principle of Digital Holography

The principle of digital holography can be explained by the principle of optical interferential recording and numerical diffraction reconstruction. First, the interferential fringe of diffracted object waves and reference waves are collected with a CCD on the hologram plane, that is, the wavefront data on the object is registered in the digital hologram. Second, the original object wavefront is reconstructed numerically from this digital hologram, based on diffraction theory with simulated reference waves and an appropriate reconstruction distance [1].

5.2.3.2 Reconstruction Principle of Computer Tomography

The reconstruction of CT is based on the principle of projection views. It can reconstruct the object's original image through the projections in different directions. The schematic map of the CT work principle is shown in Figure 5.2.6 [18].

Because of the photoelectric effect, the Compton effects, and the refraction effects when the rays pass through the object, the rays' radiation which is

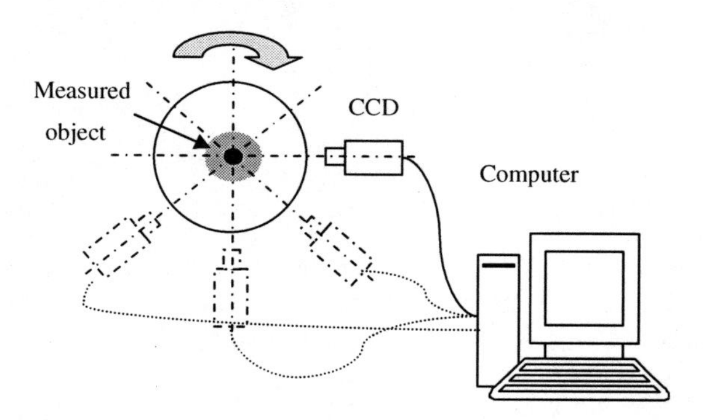

Figure 5.2.6 Schematic map of work principle of CT

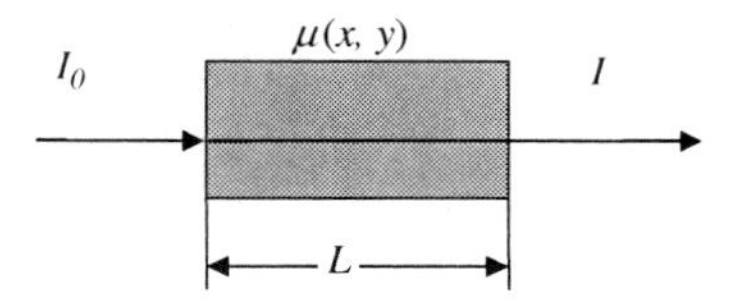

Figure 5.2.7 Ray through object with μ attenuation coefficient and length L

absorbed by the material or scattering will be weakened, as shown in Figure 5.2.7.

Attenuation follows Beer's law. When the ray with incident intensity I_0 passes through the object with attenuation coefficient μ and the length L, the incident intensity of ray passed out is I. By Beer's law, the relationships between I_0, I and $\mu(x, y)$ are:

$$I = I_0 \exp\left[-\int \mu(x, y)dxdy\right] \tag{5.2.1}$$

$$\int_L \mu(x, y)dxdy = Ln\frac{I_0}{I} \tag{5.2.2}$$

The equation in polar coordinates is expressed as:

$$\int_L \mu(x, y)dxdy = Ln\frac{I_0}{I} = p(\rho, \theta) \tag{5.2.3}$$

where the $p(s, \theta)$ is the projection values of ray passed out. By changing the projection angle θ and path p, the projection values in different directions can be obtained. Then the attenuation coefficient $\mu(x, y)$ by projection values $p(s, \theta)$ can be calculated, which is the process of reconstructing the original image using the projection images.

When using the holographic methods to register the data of the pure phase object, the optical path distance will change for the different refractive index of objects, so:

$$\phi = \phi_0 + \int_L d(x, y)dl \tag{5.2.4}$$

where the refractive index of objects is $d(x, y)$, ϕ_0 is the initial phase, ϕ is the phase of ray passed though.

Therefore, the phase or intensity projection by digital holography can be obtained, then the refractive index distribution or the structure image of related objects can be reconstructed through anti-projection.

5.2.3.3 CT Reconstruction Algorithms

The mathematical theory of CT reconstruction is based on the Radon transform which Danish scientist J. Radon proposed in 1917. Two-dimensional or three-dimensional distribution functions of the physical quantity are defined by all line integrals in their definitional domain. The two-dimensional or three-dimensional object can be expressed by a number of projections. The significance of the theory is that if all line integrals of one two-dimensional distribution function were given, then the two-dimensional distribution function can be obtained.

Now the CT reconstruction algorithms can be summarized into two ways: the transform algorithm (or the convolution algorithm) and the iterative algorithm (or the series expansion algorithm).

5.2.3.3.1 Transform Algorithm (19)

The transform algorithm is based on the Radon transform. It was the most practical reconstruction algorithm until now. The basic idea [20] is, first, collect projection images of different directions by the recording system, then do one-dimensional Fourier transforms for the projection images to get the slices in all directions. Combining all slices together, the two-dimensional Fourier transform image passing through the original point can be obtained. The last reconstruction image is acquired by the Fourier inverse transform. Using a filtering inverse projection algorithm as an example, in polar coordinates, the two-dimensional inverse Fourier transform of the reconstructed image is:

$$\begin{aligned} f(x,y) &= \int_0^{2\pi}\int_0^{\infty} F(\rho,\theta)e^{2\pi j(x\cos\theta+y\sin\theta)}\rho d\rho d\theta \\ &= \int_0^{\pi}\int_0^{\infty} F(\rho,\theta)e^{2\pi j(x\cos\theta+y\sin\theta)}|\rho|\rho d\rho d\theta \end{aligned} \tag{5.2.5}$$

where the projection of each direction is $P_\theta(\rho)$, $P_\theta(\rho)$ is its Fourier transformation.

According to the central slice theorem, Equation 5.2.5 can be transformed to:

$$f(x,y) = \int_0^{2\pi}\int_0^{\infty}\left[\int_{-\infty}^{\infty} P_\theta(\rho)e^{2\pi j\rho R}\rho d\rho\right]e^{2\pi j(x\cos\theta+y\sin\theta-R)}d\theta \tag{5.2.6}$$

where $f(x,y)$ is the image that will be reconstructed.

The principal features of the transform algorithm, which is widely used in the CT devices, are as follows: simple mathematical principles, high efficiency of reconstruction, and good quality of reconstruction with complete projection data. However, because of the inherent characteristics, it is necessary to collect the complete data through scanning the object in 180 degrees before reconstructing a certain section. Otherwise, the reconstruction result will deteriorate sharply for limited project angles or sparse projection with incomplete data. But it is hard to obtain the complete projection data in practical applications because of the testing environment, time and cost limitations.

5.2.3.3.2 Iteration Algorithm (14)

The iteration method was put forward by Kaczmarz in 1937, which is used to solve linear equations of large sparse matrices. First, assume the section that will be reconstructed as an unknown matrix, give an initial value to it, and set up a group of unknown matrix algebra equations according to the directions of projection rays. Then, compare the calculated projection values through the algebraic equation with the actual projection values, amend the unknown matrix, repeat this process three times and iterate until convergence to a fixed value is reached. Finally, obtain the image which will be reconstructed.

The typical iteration algorithm is the algebra reconstruction technique (ART) [14]. First, the rebuilt image is discrete. Then a group of initial values approximate the value of the function $f(x,y)$ through the linear combination of the basic function, as shown in Figure 5.2.8. In order to reconstruct the image, the original image was transformed discretely as the one-dimensional array $[f_1, f_2, \ldots, f_N]$, $N = n \times n$.

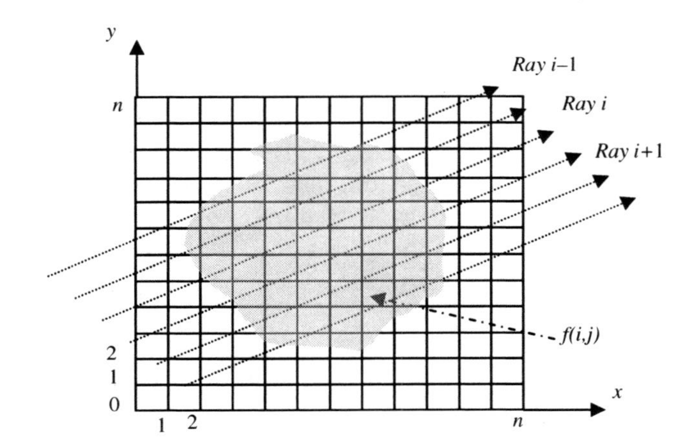

Figure 5.2.8 Reconstruction mode of ART

The process of the image projection can be expressed by Equation 5.2.7:

$$\sum_{j=1}^{N} w_{ij} f_j = p_i \tag{5.2.7}$$

where p_i is the projection value of the ray i, $i = 1, 2, \ldots, M$; w_{ij} is the weighting factor, which is the contribution of the pixel j to the line integral of the ray i; M is the number of total projection; f_j is the pixels of the box j needed.

In order to solve this problem, ART adopts the successive over-relaxation in linear algebra and the processing can be expressed by Equation 5.2.8:

$$f_j^{(k+1)} = f_j^{(k)} + \lambda \frac{p_i - \sum_{j=1}^{N} w_{ij} f_j^{(k)}}{\sum_{j=1}^{J} w_{ij}^2} w_{ij} \tag{5.2.8}$$

where k is the number of iterations; λ is the relaxation factor, usually between 0 to 2.

Therefore, it can obtain the projection value p_i of each direction and the initial value f_j^0 by testing, and realize the approximation of the function $f(x, y)$ which is needed to be rebuilt. In ART, the image rebuilding is transferred to the solved equations by algebraic iteration approximation at the beginning. And to solve the equations, for each unknown parameter, only three equations approximate its convergence limit in theory.

To solve the system of up to three equations, iterative approximation to the convergence limit for the unknown value in theory is possible.

5.2.4 Application of DHT

Traditional holographic CT is mainly applied to gas state measurement, such as the 3D measurement of a temperature field. DHT is often applied to biological morphology, such as the internal structure detection of biological tissue and the refractive index of functional materials.

5.2.4.1 Detection of Biological Tissue (8, 9)

Bio-tissue has always been a popular research topic. Florian Charrière from Switzerland first carried out the research of DHT on some objects, such as the structure of a butterfly's wing and an amoeba pollen particle. The holograms are recorded by rotating the specimen, shown in Figure 5.2.9 [8].

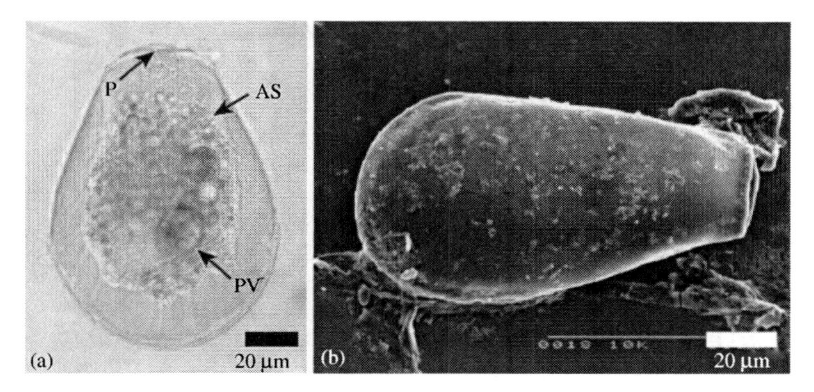

Figure 5.2.9 Images of the testate amoebae *Hyalosphenia papilio*. (a) Bright-field microscope image illustrating the amoeba itself and its content; (b) SEM image illustrating the shell (Reproduced by permission of © 2006 Optical Society of America)

Figures 5.2.9a and b are images of the testate amoebae *Hyalosphenia papilio*. Figure 5.2.9a shows a bright-field microscope image illustrating the amoeba and its content. P is a pseudostome opening through which the amoeba pseudopods emerge, AS is algal symbionts, PV is phagocyte vacuoles. Figure 5.2.9b presents an SEM image illustrating the shell.

Figures 5.2.10a and b show the cuts in the tomographic reconstructions of two different *Hyalosphenia papilios*. Discrete values of the measured refractive index n are coded in false colors, the color-coding scales being displayed on the right side of each corresponding cut.

Figure 5.2.11 shows the tomography results of a pollen cell refractive index [9]. Some cuts at different positions in the cell along the *yz*-plane and the *xz*-plane are presented in Figure 5.2.11 which shows a schematic of the presented cuts. The cuts are 2.5 μm distant from each other.

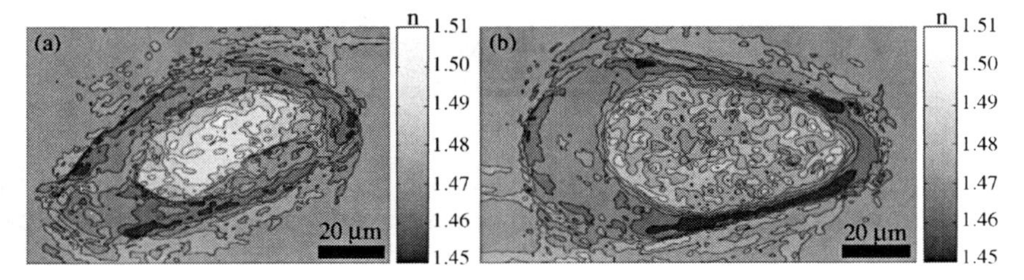

Figure 5.2.10 Cuts in the tomographic reconstructions of two different *Hyalosphenia papilio* (Reproduced by permission of © 2006 Optical Society of America)

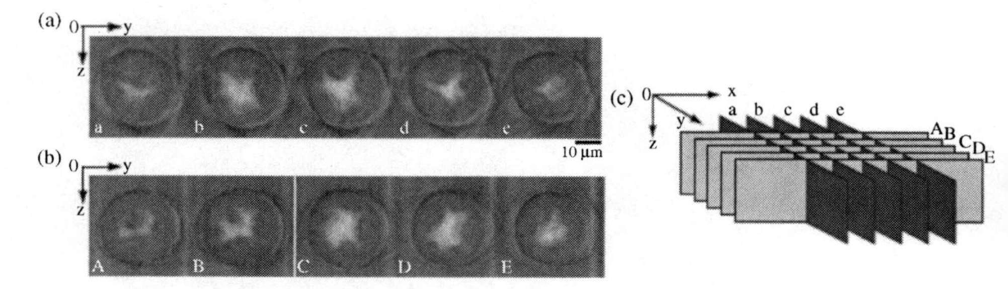

Figure 5.2.11 Cuts at different positions in the cell resp. (a) Along the *yz*-plane; (b) Along the *xz*-plane; (c) Schematic of the presented cuts (Reproduced by permission of © 2006 Optical Society of America)

5.2.4.2 Material Detection (7)

Figure 5.2.12 [7] shows tomographic images of the complex object composed of a USAF test target (part of it containing the number "2"), covered by an onion cell layer. Holograms are recorded by the optical system used to achieve OCT with a short-coherence digital holographic microscope. These images have been obtained by the reconstructions of holograms taken with coherence gates

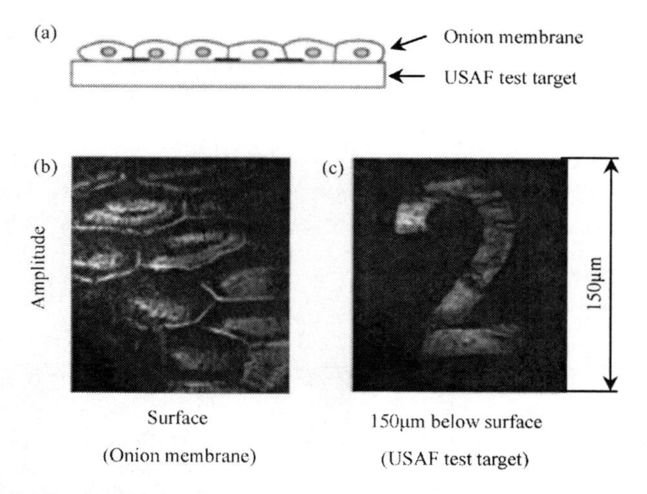

Figure 5.2.12 *En face* tomographic images of the complex object composed of a USAF test target covered by an onion cell layer. (a) Sketch of the sample; (b) *En face* image from the hologram taken with the coherence gate at the surface; (c) *En face* image from the hologram recorded 150 μm below the surface, at the level of the USAF test target (Reproduced by permission of © 2005 Optical Society of America)

at the level of the surface of the onion cells and at the level of the USAF test target. Figure 5.2.12a is the sketch of the sample, Figure 5.2.12b shows the *En face* image from the hologram taken with the coherence gate at the surface, and Figure 5.2.12c presents the *En face* image from the hologram recorded 150 μm below the surface, at the level of the USAF test target.

Shanghai University in China has also developed DHT via a few projections in 2010 [21]. The work is focused on the refractive index and a 3D structure reconstruction base on a few projections (three and four projections) for axi-symmetric and non-axisymmetric objects respectively. Some experimental results are shown in Figures 5.2.13 and 5.2.14.

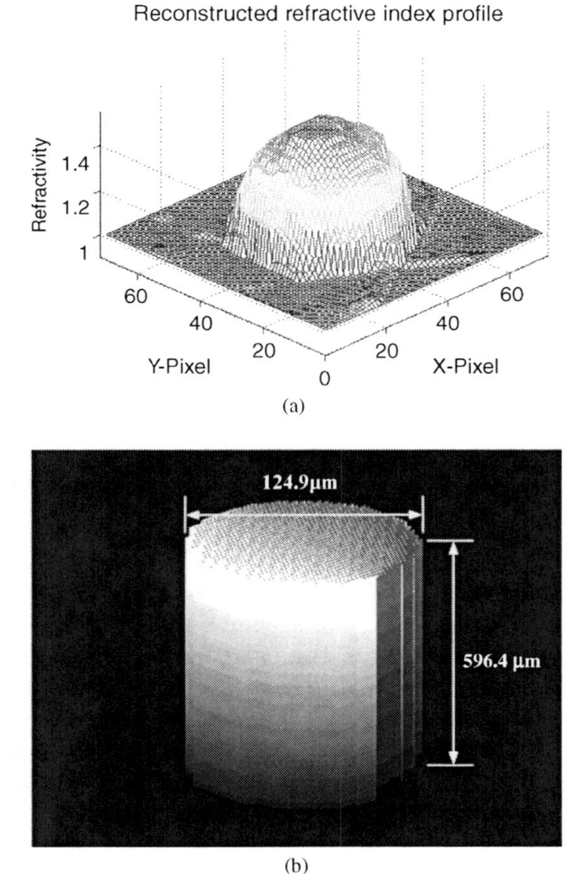

Figure 5.2.13 DHT Reconstruction result of bare fiber core using three projections. (a) 3D refractive index distribution of fiber cross-section; (b) 3D structure image of fiber by superposing of 100 slices

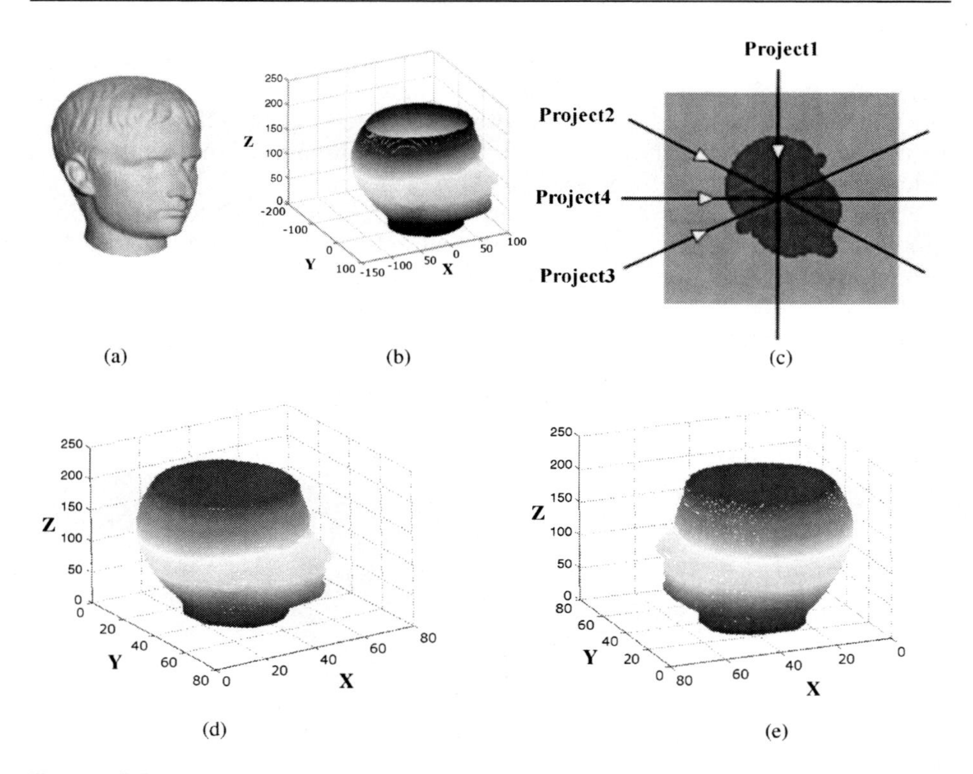

Figure 5.2.14 Reconstructed results of plaster statue using four projections based on digital holographic tomography: (a) Original model of plaster statue; (b) Data points cloud map; (c) Cross-section of No. 101 layer; (d) 3D construction image of plaster statue in view 1; (e) 3D construction image of plaster statue in view 2

The DHT reconstruction result of a section of bare fiber (an axisymmetric object) acquired by a single hologram is shown in Figure 5.2.13. Figure 5.2.13a is the 3D refractive index reconstruction of a bare fiber cross-section, which the refractive index is gradually changing from 1 to 1.5, and Figure 5.2.13b is a 3D reconstruction of the bare fiber using 100 slices superposition and the cross-section diameter of the reconstructed fiber is 122.12 μm (its actual diameter is 125 μm). During the reconstruction processing, one off-line hologram is recorded digitally and reconstructed numerically. Then, according to the axi-symmetric character of the fiber, this reconstructed phase projection is extended in three direction projections with a 60 degree interval.

The reconstruction process of a complex non-axisymmetric object (a plaster statue) using DHT is shown in Figure 5.2.14. A plaster head is used as the test sample and its original height is 220 mm in Figure 5.2.14a. The cloud map height is 212 mm in Figure 5.2.14b. In this experiment, four computer-generated holograms are obtained and reconstructed numerically. The 3D structure of the tested object is completed by using four reconstructed phase projections.

In addition, some researchers are focusing on a multi-wavelength scanning method to detect the structure or refractive index of a living cell via DHT, but have not managed better results yet.

References

1. Schnars, U. and Jŭptner, W.P.O. (2002) Digital recording and numerical reconstruction of holograms. *Meas. Sci. Technol*, **13**, 85–101.
2. Zhou, W.J., Yu, Y.J., and Asundi, A. (2009) Study on aberration suppressing methods in digital micro- holography. *Opt. Laser Eng.*, **47** (2), 264–270.
3. Zhou, W.J., Xu, Q.S., Yu, Y.J., and Asundi, A. (2009) Phase-shifting in-line digital holography on a digital micro-mirror device. *Opt. Laser Eng.*, **47** (9), 896–901.
4. Wolf, E. (1969) Three-dimensional structure determination of semi-transparent objects from holographic data. *Opt. Commun.*, **1**, 153–156.
5. Satos, K. (1986) Measurement of temperature in a flame by holographic interferometry and CT technique. Prepe Book 24th Combust Symp (in Japan), p. 16.
6. Lira, L.H. and Vest, C.M. (1987) Refraction correction in holographic interferometry and tomography of transparent objects. *Appl. Optics.*, **26** (18), 3919–3928.
7. Massatsch, P., Charrière, F., Cuche, E. *et al.* (2005) Time-domain optical coherence tomography with digital holography microscopy. *Appl. Optics.*, **44** (10), 1807–1812.
8. Charrière, F., Pavillon, N., Colomb, T. *et al.* (2006) Living specimen tomography by digital holographic microscopy: morphometry of testate amoeba. *Opt. Express*, **14** (16), 7005–7013.
9. Charrière, F., Marian, A., Montfort, F. *et al.* (2006) Cell refractive index tomography by digital holographic microscopy. *Opt. Lett.*, **31** (2), 178–180.
10. Yu, L.F. and Chen, Z.P. (2007) Digital holographic tomography based on spectral interferometry. *Opt. Lett.*, **32** (20), 3005–3007.
11. Montfort, F., Colomb, T., Charrière, F. *et al.* (2006) Submicrometer optical tomography by multiple-wavelength digital holographic microscopy. *Appl. Optics.*, **45** (32), 8209–8217.
12. Cuche, E., Poscio, P., and Depeursinge, C. (1997) Optical tomography by means of a numerical low-coherence holographic technique. *J. Opt.*, **28**, 260–264.
13. Jozwicka, A. and Kujawinska, M. (2005) Digital holographic tomography for amplitude-phase microelements. *Proc. of SPIE*, **5958**, 59580G1–59580G9.
14. Bilski, B.J., Jozwicka, A., and Kujawinska, M. (2007) 3-D phase microobject studies by means of digital holographic tomography supported by algebraic reconstruction technique. *Proc. of SPIE*, **6672**, 66720A1–66720A7.
15. Pavillon, N., Kuhn, J., Charriere, F., and Depeursinge, C. (2009) Optical tomography by digital holographic microscopy. *Proc. of SPIE*, **7371**, 7371041–7371046.
16. Sung, Y., Choi, W., Fang-Yen, C. *et al.* (2009) Optical diffraction tomography for high resolution lives cell imaging. *Opt. Express*, **17** (1), 266–277.

17. Kou, S.S. and Sheppard, C.J.R. (2008) Image formation in holographic tomography. *Opt. Lett.*, **33** (20), 2362–2364.
18. Hui, L., Xiong, W., Taoli, L. *et al.* (2007) A computed tomography reconstruction algorithm based on multipurpose optimal criterion and simulated annealing theory. *Chinese Opt. Lett.*, **5** (6), 340–343.
19. Cormack, A.M. (1963) Representation of a function by its line integrals with some radiological applications [J]. *J. Appl. Phys.*, **34**, 2722–2727.
20. Herman, G.T. (1980) *Image Reconstruction from Projections: The Fundamentals of Computerized Tomography [J]*, Academic Press, New York.
21. Wenjing, Z., Qiangsheng, X., and Yingjie, Y. (2010) Study on 3-D refractive index reconstruction based on digital holographic tomography by few projected data [J]. *Acta Phys. Sin.*, **5** (12), 8499–8511.

5.3

Digital Holographic Interferometry for Phase Distribution Measurement

Jianlin Zhao
Department of Applied Physics, School of Science, Northwestern Polytechnical University, China

5.3.1 Measurement Principle of Digital Holographic Interferometry

Digital holography is a new holographic imaging technique, in which the hologram is digitally recorded by a CCD (charge coupled device) or a CMOS (complementary metal oxide semiconductor), and the holographic image is numerically reconstructed by digitally simulating the diffraction of the hologram via a computer based on scalar diffraction theory. The digitalization of the hologram and the image reconstruction not only allow real-time holographic imaging, but also are very suitable for long-distance transmission and storage of the hologram as well as reconstruction of the holographic image. In addition, digital holography obtains the quantitative amplitude and phase data of the object wave simultaneously by employing the fast Fourier transform (FFT) and spatial filtering. Furthermore, some advanced techniques for digital image processing can be conveniently used to effectively improve the

Digital Holography for MEMS and Microsystem Metrology, First Edition. Edited by Anand Asundi.

quality of the reconstructed holographic image by eliminating aberration, noise, and other disadvantageous effects.

Digital holographic interferometry (DHI) integrates digital holography with conventional interferometry and thus can achieve fast, non-destructive, non-invasive, high resolution and full-field measurement of the phase distribution of an object wave field. With these advantages, numerous applications of DHI have been developed in many fields, such as deformation and shape analysis [1, 1–8], vibration analysis [9], microscopy [10, 10–14], refractive index and temperature distribution measurement [15, 16], particles measurement [17, 18], and so on.

5.3.1.1 Principle of Phase Measurement of the Object Wave Field

Phase measurement is one of the essential applications of the DHI. In practice, many optical measurements are based on phase measurement, such as the refractive index, density, speed, temperature distributions of transparent objects and the surface profile of reflecting objects.

Figures 5.3.1a and b show two typical experimental set-ups for recording the digital holograms of transparent and reflecting objects, respectively. In the CCD target plane (that is, the hologram recoding plane), the complex amplitudes of the object wave can be written as:

$$O(x,y) = a_o(x,y)\exp\left[j\phi_o(x,y)\right], \tag{5.3.1}$$

where, $a_o(x, y)$ and $\phi_o(x, y)$ describe the amplitude and phase distributions of the object wave, respectively. Assuming that the reference wave is a slantwise

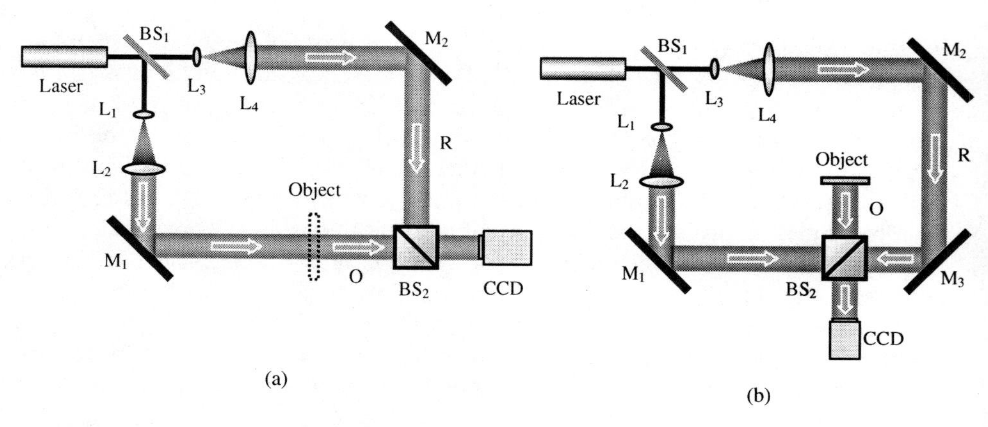

Figure 5.3.1 Typical experimental set-ups for recording digital holograms. (a) For transparent object; (b) For reflecting object
Note: L: Fourier lens; BS: beam splitter; M: mirror

collimated beam propagating in the xz plane, its complex amplitudes in the CCD target plane can be expressed as:

$$R(x, y) = a_r \exp[j2\pi\xi_r x], \tag{5.3.2}$$

where a_r and ξ_r are the amplitude and the spatial frequency of the reference wave, respectively. So the intensity distribution of the interferogram formed by superposing the object and reference waves will be expressed as:

$$\begin{aligned} I(x, y) &= |O(x, y) + R(x, y)|^2 \\ &= \left[a_o^2(x, y) + a_r^2\right] + |a_o(x, y)a_r|\exp\left[j\phi_o(x, y)\right]\exp\left[-j2\pi\xi_r x\right] \\ &\quad + |a_o(x, y)a_r|\exp[-j\phi_o(x, y)]\exp\left[j2\pi\xi_r x\right] \end{aligned} \tag{5.3.3}$$

Taking the Fourier transform of $I(x, y)$, we obtain its spatial spectrum $I(\xi, \eta)$, which is composed of three items as follows:

$$\begin{cases} I_1(\xi, \eta) = F\{a_o^2(x, y) + a_r^2\} = a_r^2\delta(\xi, \eta) + O(\xi, \eta) \otimes O(\xi, \eta) \\ I_2(\xi, \eta) = F\{|a_o(x, y)a_r|\exp\left[j\phi_o(x, y)\right]\exp(-j2\pi\xi_r x)\} = a_r O(\xi + \xi_r, \eta) \\ I_3(\xi, \eta) = F\{|a_o(x, y)a_r|\exp\left[-j\phi_o(x, y)\right]\exp(j2\pi\xi_r x)\} = a_r O(\xi - \xi_r, \eta) \end{cases}, \tag{5.3.4}$$

where $F\{\ \}$ depicts the Fourier transform operation, $O(\xi, \eta)$ is the Fourier transform of the object wave in the hologram plane, ξ and η are the spatial frequency coordinates and the symbol $\otimes$ denotes a correlation operation.

Using bandwidth filtering, the item $|I_2(\xi, \eta)|$ can be picked up from the whole spatial spectrum of the hologram. After shifting $|I_2(\xi, \eta)|$ to the origin (0, 0) in the spectrum plane, one can obtain $a_r O(\xi, \eta)$, which is just the spectrum of the object wave except for a constant a_r. So the spatial spectrum distribution of the reconstructed object wave in the object plane can be obtained according to angular spectrum theory

$$O_d(\xi, \eta) = O(\xi, \eta)\exp\left(j\frac{2\pi d}{\lambda}\right)\exp\left[-j\pi\lambda d\left(\xi^2 + \eta^2\right)\right], \tag{5.3.5}$$

where λ and d are the recording wavelength and the distance between the hologram and the object plane, respectively. Then the complex amplitude of the reconstructed object wave in the object plane can be reconstructed by:

$$O_d(x, y) = F^{-1}\{O_d(\xi, \eta)\}, \tag{5.3.6}$$

where $F^{-1}\{\ \}$ means inverse Fourier transformation. From Equation 5.3.6 we can calculate the phase distribution of the object wavefront in the object plane by:

$$\phi_d(x,y) = \arctan\left\{\frac{\mathrm{Im}[O_d(x,y)]}{\mathrm{Re}[O_d(x,y)]}\right\}. \tag{5.3.7}$$

5.3.1.2 Principle of Digital Holographic Interferometry

The complex amplitude and phase distribution of the reconstructed object wave separately obtained by Equations 5.3.6 and 5.3.7 are based on the assumption that the reference wave should be an ideally collimated beam. In practice, the laser beam could not be ideally collimated, so that the complex amplitude and phase distribution of the reconstructed object wave would be distorted due to the non-ideal collimation of the reference wave. To avoid this problem, we need to record at least two holograms to keep the reference wave invariables. One of them corresponds to the initial state of the object wave field, and the other corresponds to the changed state. For a small deformation of an object or index change of a transparent object, the change in the object wave can be considered as only in the phase distribution, while the amplitude essentially is unchanged. Taking $a_o(x, y)$ as the amplitude, and $\phi_d(x, y)$ and $\phi_d'(x, y)$ as the phase distributions of the object waves in both states before and after being changed in the object plane, respectively, the corresponding complex amplitudes of the reconstructed object waves in the object plane can be written as:

$$\begin{cases} O_d(x,y) = a_o(x,y)\exp\left[\mathrm{j}\phi_d(x,y)\right] \\ O'_d(x,y) = a_o(x,y)\exp\left[\mathrm{j}\phi'_d(x,y)\right] \end{cases}. \tag{5.3.8}$$

The phase difference $\Delta\phi(x, y) = \phi_d'(x, y) - \phi_d(x, y)$ reflects the relative change of the object field from the initial state to the changed one, which eliminates the phase distortion induced by the reference wave. According to the different processing procedures, there are three approaches [19, 20] to obtain the phase difference distribution as shown in Figure 5.3.2, that is, double-exposure, synthetic double-exposure and direct phase subtraction.

5.3.1.2.1 The Double-Exposure Method

The principle of the double-exposure DHI is similar to that of the optical double-exposure holographic interferometry. First, one can add up the intensity (gray) distributions of the two digital holograms recorded before and after

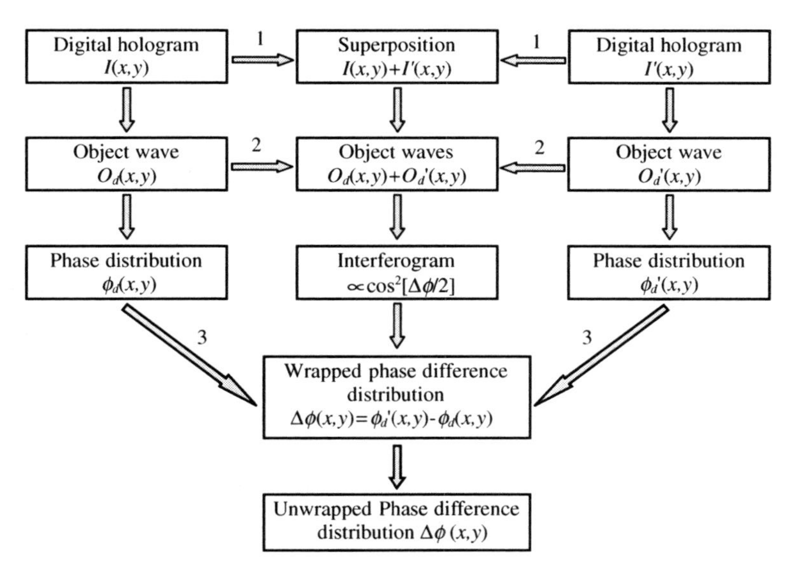

Figure 5.3.2 Approaches to obtain the phase difference distribution of the object wave field

the object field being changed to obtain a composed digital hologram, whose intensity (gray) distribution will be given by:

$$
\begin{aligned}
I(x,y)=&|O(x,y)+R(x,y)|^2+|O'(x,y)+R(x,y)|^2\\
&=\left[2a_o^2(x,y)+2a_r^2\right]+|a_o(x,y)a_r|\{\exp\left[j\phi_o(x,y)\right]+\exp\left[j\phi'_o(x,y)\right]\}\exp\left[-j2\pi\xi_r x\right]\\
&+|a_o(x,y)a_r|\{\exp\left[-j\phi_o(x,y)\right]+\exp\left[-j\phi'_o(x,y)\right]\}\exp\left[j2\pi\xi_r x\right]
\end{aligned}
\tag{5.3.9}
$$

where we assume that the reference wave is a slantwise collimated beam given by Equation 5.3.2. Similarly, taking the Fourier transform of $I(x, y)$, using bandwidth filtering to pick up the spatial spectrum component corresponding to the second item in Equation 5.3.9 and shifting it to the origin (0,0) in the spectrum plane, we obtain $a_r[O(\xi, \eta)+O'(\xi, \eta)]$, the synthetic spectrum of the two object waves in the hologram plane except for a constant a_r. Then the synthetic spectrum distribution and the complex amplitude of the two object waves in the object plane can be synchronously reconstructed according to angular spectrum theory

$$
O_d(\xi,\eta)+O'_d(\xi,\eta)=[O(\xi,\eta)+O'(\xi,\eta)]\exp\left(j\frac{2\pi d}{\lambda}\right)\exp\left[-j\pi\lambda d\left(\xi^2+\eta^2\right)\right],
\tag{5.3.10}
$$

$$O_d(x,y) + O'_d(x,y) = F^{-1}\{O_d(\xi,\eta) + O'_d(\xi,\eta)\}. \tag{5.3.11}$$

respectively. The digital superposition of the complex amplitudes gives a digital holographic interferogram reflecting the phase difference distribution of the two reconstructed object waves. The intensity distribution of the interferogram is expressed as:

$$\begin{aligned} I(x,y) &= |O_d(x,y) + O'_d(x,y)|^2 \\ &= 4a_d^2(x,y)\cos^2\left[\frac{\phi'_d(x,y) - \phi_d(x,y)}{2}\right]. \\ &= 4a_d^2(x,y)\cos^2\left[\frac{\Delta\phi(x,y)}{2}\right] \end{aligned} \tag{5.3.12}$$

Equation 5.3.12 shows that the interferogram is composed by a set of cosine-squared fringes, which reflect the tracks of the points with the same phase differences. From Equation 5.3.12 one can demodulate the phase difference $\Delta\phi(x, y) = \phi_d'(x, y) - \phi_d(x, y)$. It is interesting that Equation 5.3.12 can also be rewritten as given in [15]

$$I(x,y) = 4a_d^2(x,y)\cos^2\left[\kappa\frac{\Delta\phi(x,y)}{2}\right], \tag{5.3.13}$$

where κ is an integer which will cause an increase of κ times of the number of the interference fringes. That means each fringe will be divided as κ stripes, so that some details can be more clearly observed.

5.3.1.2.2 The Synthetic Double-Exposure Method

The principle of the synthetic double-exposure DHI is similar to that of the optical synthetic double-exposure holographic interferometry. First, one can separately reconstruct the complex amplitudes of the two object waves according to the procedure presented in Section 5.3.1.1, and then by calculating their superposition intensity a similar result as shown in Equations 5.3.12 or 5.3.13 can also be obtained.

5.3.1.2.3 The Direct Phase Subtraction Method

The phase subtraction method is a particular method of digital holographic interferometry. As mentioned in Section 5.3.1.1, one can directly pick up the phase distributions from the complex amplitudes of the reconstructed object waves in different states and then calculate the relative phase difference by direct subtraction of the phase distribution functions of the reconstructed object waves in different states, $\Delta\phi(x, y) = \phi_d'(x, y) - \phi_d(x, y)$.

In approach 1, the holographic interferogram can be obtained only by one numerical reconstruction operation, while in approaches 2 and 3 it require two numerical reconstruction operations to obtain the holographic interferogram or wrapped phase difference distribution. To observe the variation process of a dynamic object field, we need to continually record multi-frames of the holograms in different states, and then by comparing the reconstructed object wave in each state with that in the initial one, we can obtain the relative variations in different states or a video reflecting the dynamic variation process of the object field. For this purpose, approach 2 or 3 is more convenient. Furthermore, for a phase object, the phase variation of the object wave directly represents the variation of the object field, so to obtain the absolute phase difference, approach 2 or 3 should be the first choice.

It is also noted that the holographic interferogram obtained by double exposure or synthetic double exposure methods and the pattern of the relative phase difference distribution obtained by phase subtraction method are all wrapped, in which the phase or phase difference values are confined in a $\pm\pi$ scale. Thus, it is necessary to unwrap the phase or phase difference distribution to obtain the true phase or phase difference value. According to the practical requirement, many appropriate and effective unwrapping algorithms can be selected and applied to obtain the true phase distribution of the object wave field.

5.3.2 Applications of Digital Holographic Interferometry in Surface Profile Testing of MEMS/MOEMS

The property characterization and defect testing of the micro-electromechanical system (MEMS) or micro-opto-electromechanical system (MOEMS) are one of the most important steps in their manufacture and applications. Usually, three typical apparatuses are used to test MEMS/MOEMS, which are: (1) the scanning electronic microscope (SEM) [21]; (2) the confocal laser scanning microscope (CLSM) [22–24]; and (3) the white-light interference microscope (WLIM) [25–27]. The problem is that all of these have mechanical scanning systems which are not suitable for the dynamic surface deformation of the microstructures. Other apparatuses including a stylus profilometer, or an atomic force microscope (AFM) also have the same problem.

DHI has been proved to have the ability to overcome the above shortcomings [28–31]. Figure 5.3.3 shows an experimental set-up for testing MOEMS/MEMS surface deformation based on DHI [32]. A thin laser beam is divided

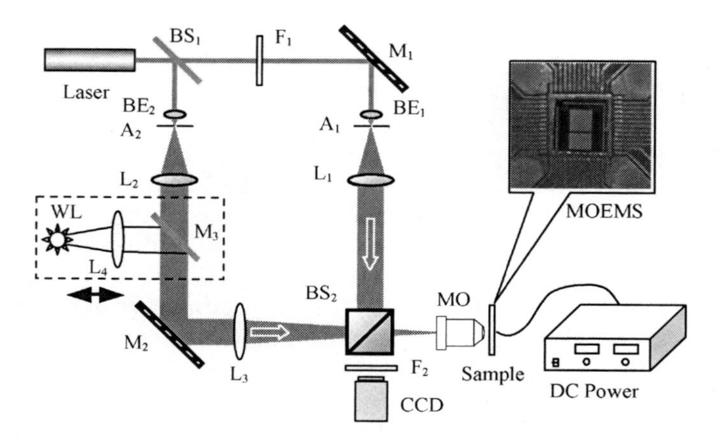

Figure 5.3.3 Experimental set-up for surface deformation testing of MOEMS/MEMS based on DHI

Note: L: lens; BS: beam splitter; M: mirror; BE: beam expander; A: pinhole; F: neutral filter; WL: white light source; MO microscope objective

into two parts by a beam splitter BS_1. One beam is reflected by mirror M_1 and goes through the beam expander BE_1, pinhole A_1 and lens L_1 to form the collimated reference beam. The other is also converted to an expanded and collimated beam via beam expander BE_2, pinhole A_2 and lens L_2, then illuminates the sample through a reversed telescope system formed by convex lens L_3 and microscope objective MO. The reference beam and the object beam scattered from the surface of the sample and magnified by MO are collected via the beam splitter BS_2 and form the interferogram on the CCD target. Two neutral filters F_1 and F_2 are used to attenuate the beam intensity. A white-light source (WL in the rectangle box of the figure) is used to get an in-focus white-light image before deformation of the sample to make the template without a diffraction effect and determine the cantilever beam's edge.

After numerical reconstruction, we can get two reconstructed object wavefronts, $O_1 = O\exp(i\phi_1)$ (for the initial state) and $O_2 = O\exp(i\phi_2)$ (for the changed state), respectively. Considering a reflection configuration, the phase difference $\Delta\phi$ between the two reconstructed object wavefronts is given by:

$$\Delta\phi = \phi_2 - \phi_1 = \frac{4\pi}{\lambda}\Delta L, \tag{5.3.14}$$

where ΔL is the optical path difference, $\Delta\phi$ or ΔL reflects the surface deformation of the MOEMS/MEMS. By phase subtraction between the object wavefronts reconstructed from the holograms recorded in different changed states

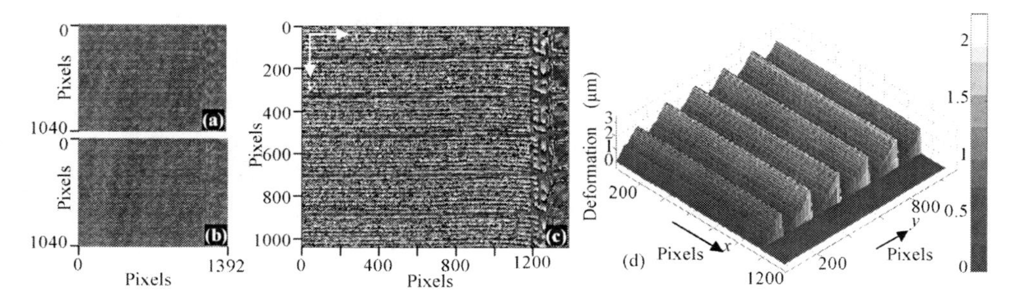

Figure 5.3.4 Measurement result on the surface deformation of the micro cantilever beam array (32). (a) and (b) Holograms recorded at the drive voltage of 0 V and 130 V, respectively; (c) Wrapped phase pattern from (a) and (b); (d) Phase unwrapped result (off-plane displacements of the cantilever beam surface) (Reproduced by permission of ©2009 Optical Society of America)

and that recorded at the initial state, a set of the wrapped phase difference distributions will be obtained.

Figure 5.3.4 shows the deformation measurement result of a micro cantilever beam array driven with voltage [32]. Where, Figure 5.3.4a and b are the holograms recorded at the drive voltage of 0 V and 130 V, respectively; Figure 5.3.4c is the corresponding wrapped phase difference distribution between two reconstructed object wavefronts from Figure 5.3.4a and b, and Figure 5.3.4d gives the phase unwrapping result, that is the off-plane displacements of the cantilever beam surface. It shows that DHI provides an efficient and accurate way for the surface deformation testing of MOEMS/MEMS and can be readily applied in designing and fabricating reliable MOEMS/MEMS.

5.3.3 Applications of Digital Holographic Interferometry in Measuring Refractive Index Distribution

For transparent objects, assuming that a collimated light beam passes the object along z direction and the refractive index distributions of the object before and after being changed are $n(x, y, z)$ and $n'(x, y, z)$, respectively, the relationship of the phase variation $\Delta\phi(x, y)$ with refractive index change can be given by [33]

$$\Delta\phi(x,y) = \frac{2\pi}{\lambda}\int_0^L [n'(x,y,z) - n(x,y,z)]\mathrm{d}z = \frac{2\pi}{\lambda}\int_0^L \Delta n(x,y,z)\mathrm{d}z, \qquad (5.3.15)$$

where, λ is the wavelength and L is the geometrical path length along the propagation direction. If the refractive index distribution is uniform in z direction, from Equation 5.3.15 we can obtain the refractive index change in the transparent object

$$\Delta n(x, y) = n'(x, y) - n(x, y) = \frac{\lambda}{2\pi L} \Delta\phi(x, y). \tag{5.3.16}$$

Equation 5.3.16 shows that the phase difference distribution directly reflects the variation of the refractive index distribution in the object when L is a constant. For transparent flow fields, such as acoustic standing wave field in air, temperature field of air, plasma plume, water flow field, and so on, the refractive index distribution may also represent the density or even speed distributions of the flow fields. That means digital holographic interferometry also presents an effective solution to visually measure the changes or fluctuations of the density or speed distributions in such transparent objects.

5.3.3.1 Measurement of Light-Induced Index Change in Photorefractive Crystals

Optical induction technique is one of the recognized effective ways for fabricating photonic structures in some bulk photorefractive crystals. In such materials, photonic structures (for example, optical waveguides or waveguide arrays) can be created solely by light illuminations at very low power levels (microwatt or even lower), and can manipulate intense probe beams at less photosensitive wavelengths. Moreover, the induced structures can be dynamically controlled, which may be important for applications of adaptive interconnections. The index distribution is one of the key parameters of the light-induced photonic structures, and is closely related to the fabrication technique and the characteristics for light manipulation.

The traditional means of measuring the light-induced index changes in photorefractive crystals is mainly interferometry [34–36]. Digital holographic interferometry provides a more convenient and efficient method of visually measuring of the refractive index distributions of the light-induced photonic structures [16]. The experimental set-up is generally made up of a Mach–Zehnder interferometer as shown in Figure 5.3.5. A thin laser beam is expanded and collimated through a reversed telescope RT, and split into two beams. Lenses L_1 and L_3, L_2 and L_3 form two 4f-systems, respectively. The rear face of a photorefractive crystal is placed at the front focal plane of lens L_1. The target plane of a CCD with high resolution and high sensitivity is placed at the rear focal plane of lens L_3 to record the image holograms. The beam propagating through the crystal sample in the Mach–Zehnder interferometer acts as the object beam, and the other as the reference beam. The polarizations of the two beams are always kept parallel to the c axis of the crystal to probe the extraordinary index changes. To avoid the probe beam producing additional index changes in the crystal, the output power density of the laser beam is adjusted

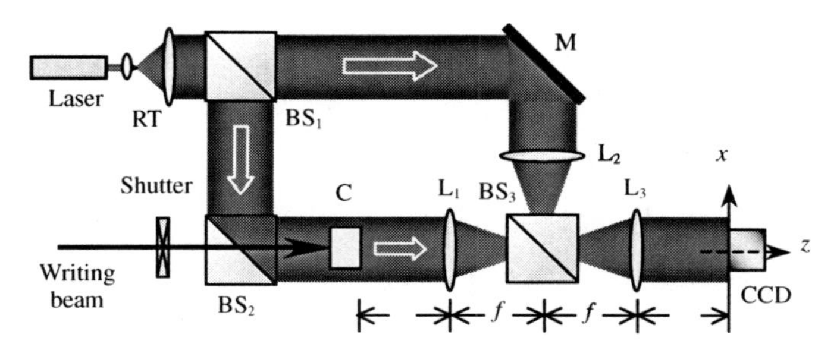

Figure 5.3.5 Experimental set-up for visualization measurement of the index distributions of the light-induced photonic structures (*Source*: (16)) (Reproduced by permission of © 2003 Chinese Physics Society)
Note: RT: inverted telescope; BS: beam splitter; L: lens

to a very low level. A writing beam is employed to irradiate the photorefractive crystals to induce the photonic structures. To determine the refractive index profiles of the induced structures, two holograms need to be recorded before and after the crystal is irradiated.

5.3.3.1.1 Index Change Induced by a Gaussian Beam

As a experimental demonstration, a circular Gaussian beam with a polarization perpendicular to the crystal *c* axis is introduced to irradiate a single poled *y*-cut $LiNbO_3$:Fe crystal. Figure 5.3.6 shows the experimental results, where Figures 5.3.6a and b are the holograms recorded by CCD before and after irradiation, and Figures 5.3.6c and d display the corresponding wrapped phase distributions of the object wavefonts reconstructed from the holograms in Figures 5.3.6a and b, respectively; Figures 5.3.6e and f separately depict the wrapped and unwrapped phase difference distributions between Figures 5.3.6c and d; Figure 5.3.6g displays the index changes along the white lines in Figure 5.3.6f, and (h) is the corresponding two-dimensional distribution of the index changes. Figure 5.3.7 show the simulation results of the Gaussian beam induced refractive index changes. It is obvious that the experimental results in Figure 5.3.6 are coincident with the simulation results in Figure 5.3.7.

5.3.3.1.2 Relative Index Profile of Light-Induced Waveguide Array (37, 38)

Figure 5.3.8 shows an improvement on the experimental set-up in Figure 5.3.5 for writing one- or two-dimensional waveguide arrays in photorefractive crystal and measuring the relative index profile of the light-induced waveguide array. A collimated coherent beam propagates through a Fresnel's biprism BP_1

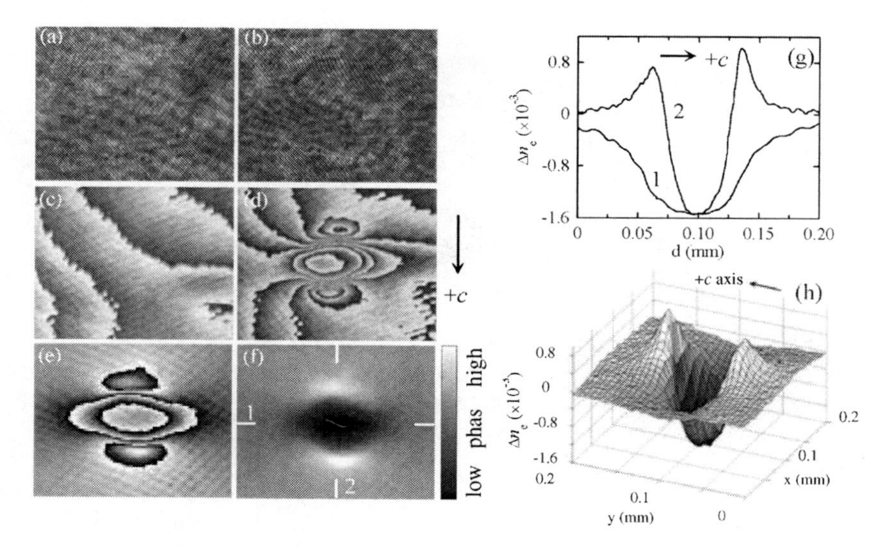

Figure 5.3.6 Measurement results of the Gaussian beam induced refractive index changes in $LiNbO_3$:Fe crystal (16). (a) and (b) Holograms before and after the crystal being illuminated; (c) and (d) Wrapped phase distributions of the object wavefonts reconstructed from (a) and (b); (e) and (f) Wrapped and unwrapped phase difference distributions between (c) and (d); (g) Index changes along the white lines in (f); (h) Two-dimensional distribution of the index changes (Reproduced by permission of © 2003 Chinese Physics Society)

or a pair of Fresnel's biprisms BP_1 and BP_2 to form two- or multi-beam interference fields as the writing beam. The crystal located in the interference field will record the intensity pattern in terms of index change due to the photorefractive effect. As a result, a waveguide array will be created in the crystal bulk. To obtain a larger index modulation, the crystal c axis is adjusted to be parallel to the grating vector of the interference field as soon as possible.

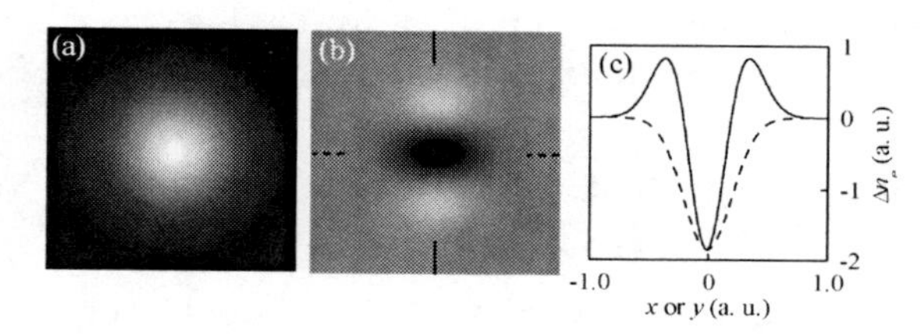

Figure 5.3.7 Simulation results of the Gaussian beam induced refractive index changes in $LiNbO_3$:Fe crystal. (a) Intensity distribution of the writing beam; (b) Index change in terms of gray; (c) Index changes along the lines in (b)

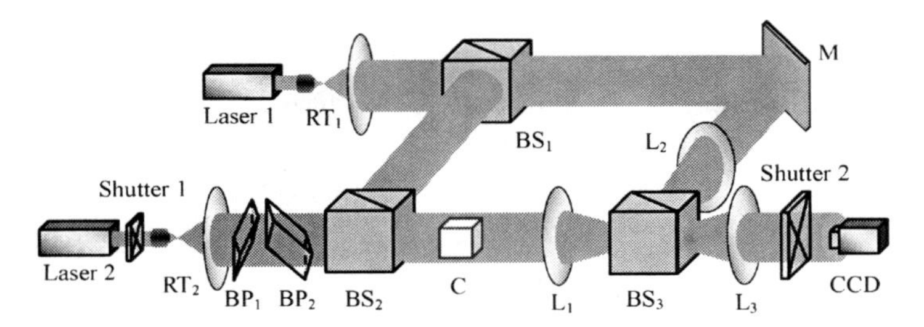

Figure 5.3.8 Experimental set-up for fabricating a light-induced waveguide array in a photorefractive crystal employing two- or multi-beam interference fields and measuring the relative index profile of the waveguide array
Note: RT: reversed telescope; BS: beam splitter; BP: Fresnel's biprism; L: lens

Figure 5.3.9 shows the measurement results of a planar waveguide array induced in $LiNbO_3$:Fe crystal with illumination of two-beam interference field formed by Fresnel's biprism, where Figures 5.3.9a and b depict the holograms before and after irradiation, respectively, Figure 5.3.9c is the near-field pattern of the waveguide array, and Figures 5.3.9d–f are the relative index distributions of the planar waveguide array measured by using DHI.

To induce a two-dimensional channel waveguide array in a photorefractive crystal, we can irradiate the crystal twice with the two-beam interference field by rotating the single Fresnel's biprism, or irradiate the crystal once with a four-beam interference field formed by a pair of identical Fresnel's biprisms BP_1 and BP_2 with different orientations. The orientations of BP_1 and BP_2 can be

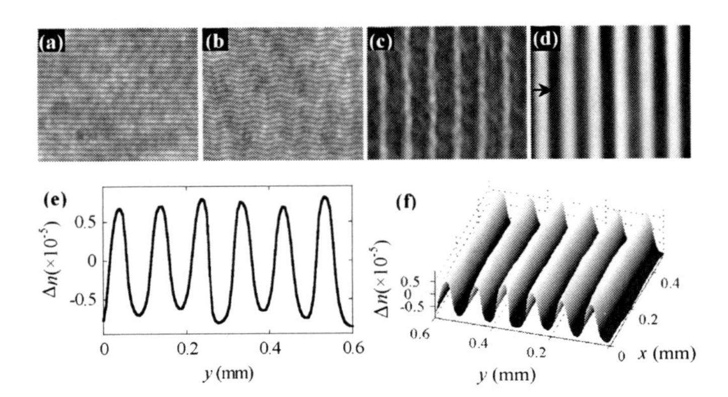

Figure 5.3.9 Measurement results of the relative index distribution of a planar waveguide array induced in $LiNbO_3$:Fe crystal. (a) and (b) Holograms recorded before and after the illumination; (c) Near-field pattern of the waveguide array; (d)–(f) The relative index distribution, the index profile along the arrow in (d) and the three-dimensional display of (d)

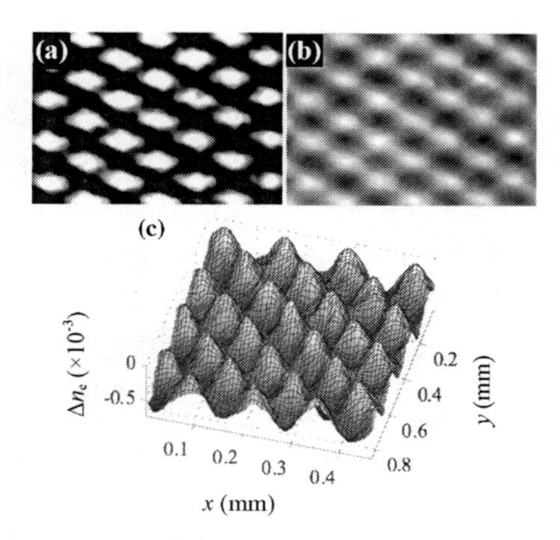

Figure 5.3.10 Measurement result of the channel waveguide array induced in $LiNbO_3$:Fe crystal by double irradiation with the two-beam interference field (37). (a) Near-field pattern of the waveguide array; (b) and (c) The relative index distributions in two- and three-dimensions (Reproduced by permission of © 2004 Chinese Physics Society)

rotated in the transverse plane. Figure 5.3.10 shows the measurement result of the channel waveguide array induced in $LiNbO_3$:Fe crystal by double irradiation with the two-beam interference field [37], where Figure 5.3.10a is the near-field pattern of the waveguide array, and Figures 5.3.10b and c are the

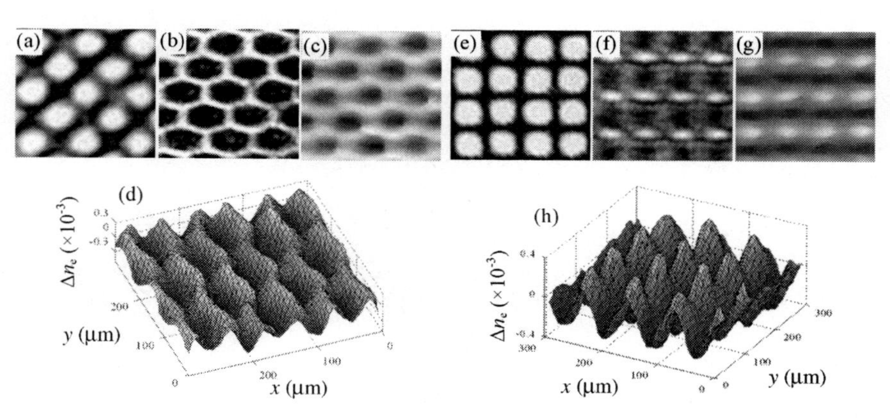

Figure 5.3.11 Measurement result of the channel waveguide arrays induced in $LiNbO_3$:Fe crystal by four-beam interference fields. (a) and (e) Intensity distributions of writing beams; (b) and (f) Near field patterns of the waveguide arrays; (c) and (g) Relative index distributions of the waveguide arrays; (d) and (h) Three-dimensional displays of (c) and (g), respectively

relative index distributions in two- or three-dimensions of the channel waveguide array measured by using DHI, respectively.

Figure 5.3.11 shows the measurement results of the channel waveguide array induced in $LiNbO_3$:Fe crystal by four-beam interference field in different configurations, where Figures 5.3.11a and e describe the intensity patterns of the interference fields illuminating the crystal, Figures 5.3.11b and f are the near-field patterns of the induced waveguide arrays, Figures 5.3.11c and g are the measured index distributions in terms of gray, and Figures 5.3.11d and h are the three-dimensional display of (c) and (g), respectively.

5.3.3.2 Measurement of Acoustic Standing Wave Field

The acoustic standing wave with high intensity can produce an intensive pressure, namely acoustic radiation pressure. Under certain conditions the acoustic radiation pressure can levitate an object with density many times more than atmosphere in the air [39]. The levitated object is located at one of the standing wave nodes. The node number is identical to that of the resonant modes [40]. Considering that the acoustic field is invisible and the traditional direct detection would disturb the actual distribution of the acoustic radiation pressure, as a visible, nondestructive and full-field measurement method, DHI can be used to solve this problem [41, 42]. Figure 5.3.12 shows the measurement set-ups

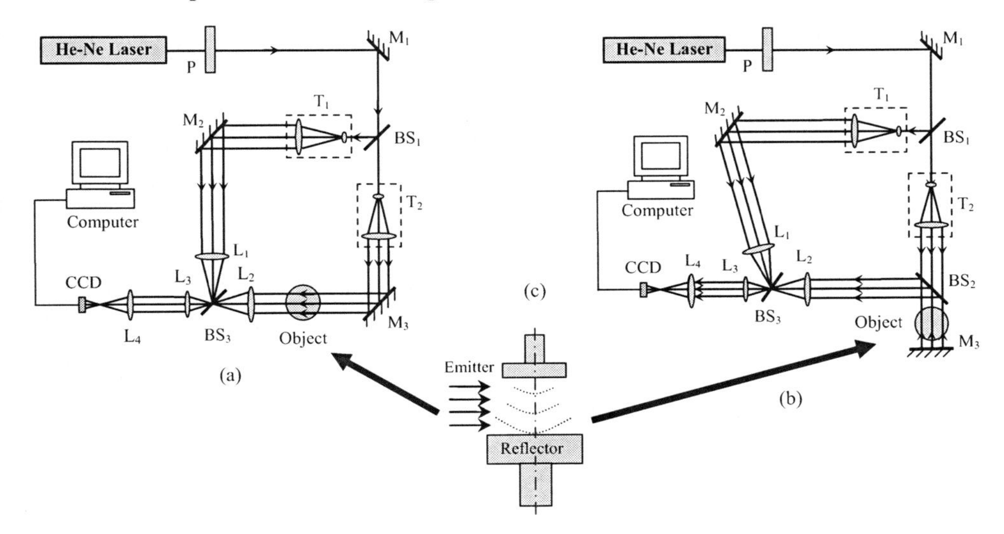

Figure 5.3.12 Experimental set-ups for measuring the acoustic radiation pressure distribution based on DHI (42). (a) Basic set-up; (b) Set-up with phase multiplication; (c) Acoustic levitator

Note: P: polarizer; M: mirror; L: lens; T: reversed telescope; BS: beam splitter (Reproduced by permission of 2009 Elsevier)

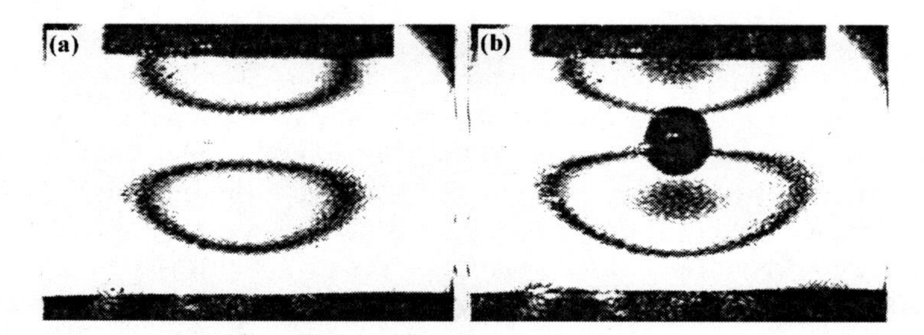

Figure 5.3.13 Digital holographic interferograms of the second resonant mode at the same inspiriting current (42). (a) Without levitated object; (b) With levitated object (Reproduced by permission of 2009 Elsevier)

[42], where Figure 5.3.12a is a basic set-up based on a Mach–Zehnder interferometer and Figure 5.3.12b is an improved one with phase multiplication [42]. As shown in Figure 5.3.12a, a thin He-Ne laser passes through a polarizer P and is divided by a beam splitter BS_1. The reflected beam is expanded and collimated by reversed telescope T_1 as the reference beam after passing though the lens group L_1, L_3 and L_4. The transmitted beam is expanded and collimated by reversed telescope T_2 and then passes through the acoustic levitator as the object beam after passing through the lens group L_2, L_3 and L_4 and finally interferes with the reference beam on the CCD target. By adjusting the distances between CCD, lens L_4 and the acoustic levitator, an image hologram can be received by the CCD. Different from the set-up in Figure 5.3.12a, the object beam passes through the acoustic levitator twice by use of the beam splitter BS_2 in Figure 5.3.12b which is helpful to double the phase variation of the object beam and increase the measurement sensitivity. Here mirror M_3 must be absolutely perpendicular to the incident beam to make the reflected beam and the incident beam superposable.

Figure 5.3.13 depicts the reconstructed digital holographic interferograms [42] of the second resonant mode with and without a levitated object at a certain inspiriting current with the set-up shown in Figure 5.3.12b. It is obvious that the fringe patterns clearly display the acoustic radiation pressure distributions and the fringe number in Figure 5.3.13b is more than that in Figure 5.3.13a, which means the acoustic radiation pressure has a slight increase when there is a levitated spherule in the acoustic standing wave field.

5.3.3.3 Measurement of Plasma Plume Field

The distribution of the invisible plasma plume is currently an important research topic for a microwave plasma thruster (MPT). DHI offers an effective, visible and noninvasive approach to measure the electronic density of the MPT. The plasma plume of MPT can be measured by using a set-up similar to

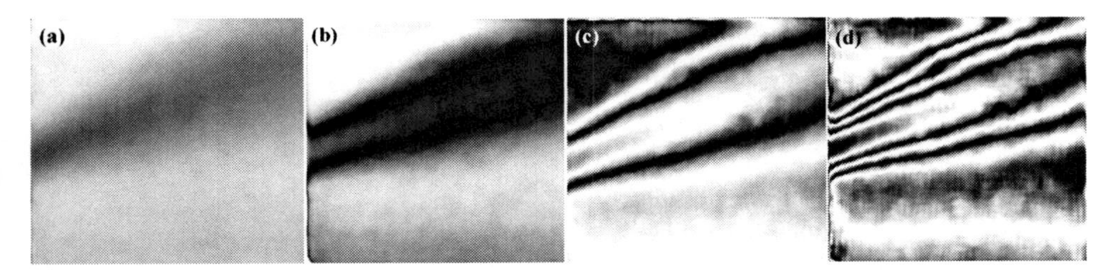

Figure 5.3.14 Digital holographic interferograms reflecting the distribution of the MPT plasma plume (43). (a) Original interferogram; (b) Multiplicated phase deference ($\kappa = 4$); (c) Multiplicated phase deference ($\kappa = 10$); (d) Multiplicated phase deference ($\kappa = 20$) (Reproduced by permission of © 2005 Acta Photonica Sinica)

Figure 5.3.12a. It need to record two digital holograms with and without the plasma plume. Figure 5.3.14 shows the measurement results [43], where Figure 5.3.14a is the original digital holographic interferogram reflecting the distribution of the electronic number of the MPT plasma plume; Figures 5.3.14b–d are the same interferogram but with different multiplications of phase deference.

5.3.3.4 Measurement of Temperature Distribution in Air Field

DHI provides an effective approach to measure the temperature distribution of a transparent object such as an air field. An experimental set-up similar to Figure 5.3.12a can be also used to record the hologram of the temperature field in air field. Figure 5.3.15 shows the measurement results of the temperature distribution around an electric iron in the air [15], where Figures 5.3.15a and c are the digital holographic interferograms reflecting the temperature distribution around the head and the wall of the electric iron, respectively; Figures 5.3.15b

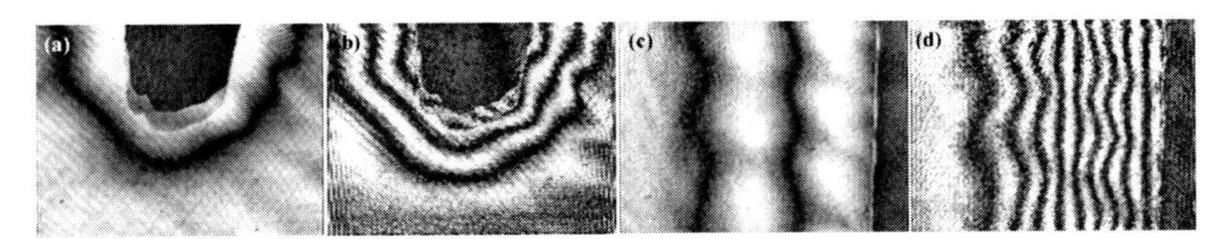

Figure 5.3.15 Digital holographic interferograms reflecting the temperature distribution. (a) Holographic interference patterns of the temperature field around the head of an electric iron (15); (b) Multiplicated phase deference of (a) ($\kappa = 3$); (c) Holographic interference patterns of the temperature field around the wall of the electric iron; (d) Multiplicated phase deference of (c) ($\kappa = 4$) (Reproduced by permission of © 2002 Acta Photonica Sinica)

and d are the same interferograms but with different multiplications of phase deference.

5.3.3.5 Visualization Measurement of Turbulent Flow Field in Water

In fluid dynamics research, flow visualization is the most intuitive instrumentality for cognizing the flow field distribution. Traditional optical methods for flow visualization, such as the Schlieren technique [44] and the shadowgraph technique [45], are restricted in their practical application due to the available test section size and reduced sensitivity. Besides, optical velocity methods based on tracking the position and velocity of the tracer particles in a liquid, such as particle image velocimetry, or laser-Doppler velocimetry [46] also have the problem that sometimes the tracer particles cannot follow the fluid quite well.

DHI offers an advanced technique to visualize the flow field [47–52]. Most flows are complex, containing numerous vortices with different sizes and periodicities. Among them the turbulent Karman vortex street flowing past circular cylinders involves most of the characteristic features of technical applications, so it is an ideal test case for further investigating the application of digital holography on measuring more complex flow.

Figure 5.3.16 shows the experimental set-up for measuring the Karman vortex streets formed behind a circular cylinder in water channel and its evolution [53]. A collimated laser beam vertically propagates through the flow field to form the object beam, and a telecentric lens set (TL) is used to image a large field of view without distortion. The water channel is a transparent rectangular

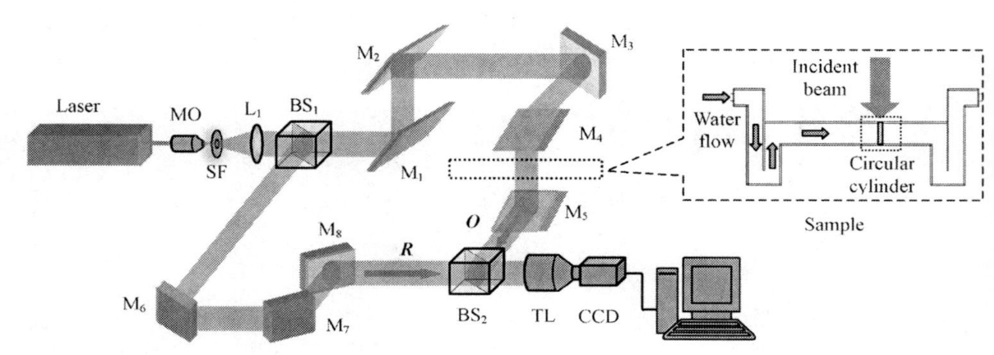

Figure 5.3.16 Experimental set-up for measuring the Karman vortex streets formed behind a circular cylinder in a water channel and its evolution (*Source*: [53]) (Reproduced by permission of © 2005 Optical Society of America)
Note: BS: beam splitter; M: mirror; TL: telecentric lens set; L: lens; MO: microscope objective; SF: pinhole

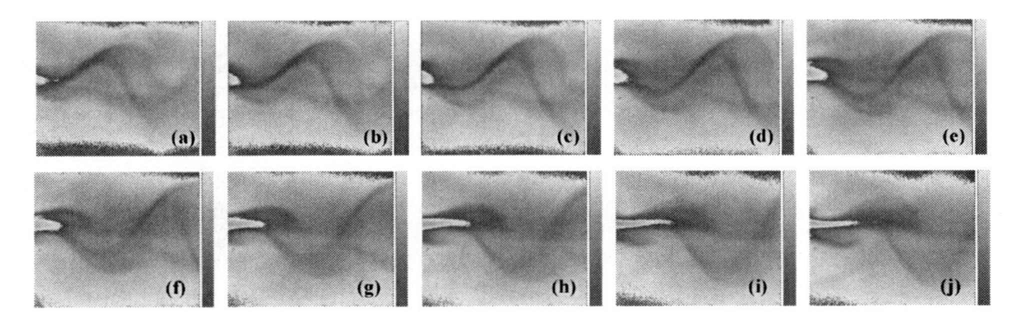

Figure 5.3.17 Reconstructed two-dimensional wrapped phase difference distributions of the Karman vortex street. (a)–(j) Ten frames of the movie in a period (*Source*: (33)) (Reproduced by permission of Society of Photo-optical Instrumentation Engineers © 2009)

container. The circular cylinder with a smooth surface is made of metal and vertically inserted between the top and bottom walls of the channel.

First, a hologram in a still state and following holograms in a steady flow with a certain velocity are recorded and numerically reconstructed, respectively. Then, a series of sequential phase difference distributions of the object wave in a steady flow state relative to that in the still state can be obtained. Figure 5.3.17 shows a sequence of the reconstructed two-dimensional wrapped phase difference distributions [53], where each picture separately presents a frame of the Karman vortex streets in one period. From Figure 5.3.17, the characteristic of the periodic shedding of the vortices can be clearly seen, and in the region close to the channel wall, the refractive index of the water presents an evident gradient and random fluctuation. We can also calculate the periodicity in that the same distribution of the Karman vortex street reoccurs.

References

1. Schnars, U. and Jüptner, W. (1994) Digital recording and reconstruction of holograms in hologram interferometry and shearography. *Appl. Opt.*, **33** (20), 4373–4377.
2. Nilsson, B. and Carlsson, T.E. (1998) Direct three-dimensional shape measurement by digital light-in-flight holography. *Appl. Opt.*, **37** (34), 7954–7959.
3. Pedrini, G., Froening, P., Fessler, H., and Tiziani, H.J. (1998) In-line digital holographic interferometry. *Appl. Opt.*, **37** (26), 6262–6269.
4. Kebbel, V., Adams, M., Hartmann, H.J., and Jüptner, W. (1999) Digital holography as a versatile optical diagnostic method for microgravity experiments. *Meas. Sci. Technol.*, **10** (10), 893–899.
5. Santoyo, F.M., Pedrini, G., Schedin, S., and Tiziani, H.J. (1999) 3D displacement measurements of vibrating objects with multi-pulse digital holography. *Meas. Sci. Technol.*, **10** (12), 1305–1308.

6. Wagner, C., Seebacher, S., Osten, W., and Jüptner, W. (1999) Digital recording and numerical reconstruction of lensless Fourier holograms in optical metrology. *Appl. Opt.*, **38** (22), 4812–4820.
7. Schedin, S., Pedrini, G., and Tizian, H.J. (2000) Pulsed digital holography for deformation measurements on biological tissues. *Appl. Opt.*, **39** (16), 2853–2857.
8. Seebacher, S., Osten, W., Baumbach, T., and Jüptner, W. (2001) The determination of material parameters of microcomponents using digital holography. *Opt. Laser Eng.*, **36** (2), 103–126.
9. Pedrini, G., Osten, W., and Gusev, M.E. (2006) High-speed digital holographic interferometry for vibration measurement. *Appl. Opt.*, **45** (15), 3456–3462.
10. Carl, D., Kemper, B., Wernicke, G., and Von Bally, G. (2004) Parameter-optimized digital holographic microscope for high-resolution living-cell analysis. *Appl. Opt.*, **43** (36), 6536–6544.
11. Guo, P. and Devaney, A.J. (2004) Digital microscopy using phase-shifting digital holography with two reference waves. *Opt. Lett.*, **29** (8), 857–859.
12. Ferraro, P., Grilli, S., and Alfieri, D. (2005) Extended focused image in microscopy by digital holography. *Opt. Express*, **13** (18), 6738–6749.
13. Mann, C.J., Yu, L., Lo, C., and Kim, M.K. (2005) High-resolution quantitative phase-contrast microscopy by digital holography. *Opt. Express*, **13** (22), 8693–8698.
14. Cuche, E., Marquet, P., and Depeursinge, C. (1999) Simultaneous amplitude-contrast and quantitative phase-contrast microscopy by numerical reconstruction of Fresnel off-axis holograms. *Appl. Opt.*, **38** (34), 6994–7001.
15. Zhao, J.L. and Tan, H.Y. (2002) Measuring three-dimensional temperature field by digital holographic interferometry. *Acta, Opt. Sin.*, **22** (12), 1447–1451 (in Chinese).
16. Zhao, J.L., Zhang, P., Zhou, J.B. *et al.* (2003) Visualizations of light-induced refractive index changes in photorefractive crystals employing digital holography. *Chin. Phys. Lett.*, **20** (10), 1748–1751.
17. Murata, S. and Yasuda, N. (2000) Potential of digital holography in particle measurement. *Opt. Laser Technol.*, **32** (7–8), 567–574.
18. Lebrun, D., Benkouider, A.M., Coetmellec, S., and Malek, M. (2003) Particle field digital holographic reconstruction in arbitrary tilted planes. *Opt. Express*, **11** (3), 224–229.
19. Schnars, U. (1994) Direct phase determination in hologram interferometry with use of digitally recorded holograms. *J. Opt. Soc. Am. A.*, **11** (3), 2011–2015.
20. Zhao, J.L. (2010) Information optics, in *Handbook of Optics* (ed. D.Z. Li), Shaanxi Science and Technology Publishing House (in Chinese).
21. Novak, E. (2005) MEMS metrology techniques. *Proc. SPIE*, **5716**, 173–181.
22. Maitland, K.C., Shin, H.J., Ra, H. *et al.* (2006) Single fiber confocal microscope with a two-axis grimbaled MEMS scanner for cellular imaging. *Opt. Express*, **14** (19), 8604–8612.
23. Li, Z., Herrmann, K., and Pohlenz, F. (2007) Lateral scanning confocal microscopy for the determination of in-plane displacements of microelectromechanical systems devices. *Opt. Lett.*, **32** (12), 1743–1745.
24. Tiziani, H.J., Achi, R., Kramer, R.N., and Wiegers, L. (1996) Theoretical analysis of confocal microscopy with microlenses. *Appl. Opt.*, **35** (1), 120–125.

25. Roy, M., Sheppard, C.J.R., Cox, G., and Hariharan, P. (2006) White-light interference microscopy: a way to obtain high lateral resolution over an extended range of heights. *Opt. Express*, **14** (15), 6788–6793.
26. Schmit, J. and Olszak, A. (2002) High-precision shape measurement by white-light interferometry with real-time scanner error correction. *Appl. Opt.*, **41** (28), 5943–5950.
27. Pavlicek, P. and Soubusta, J. (2003) Theoretical measurement uncertainty of white-light interferometry on rough surfaces. *Appl. Opt.*, **42** (10), 1809–1813.
28. Coppola, G., Ferraro, P., Iodice, M. *et al.* (2004) A digital holographic microscope for complete characterization of microelectromechanical systems. *Meas. Sci. Technol.*, **15** (3), 529–539.
29. Ferraro, P., De Nicola, S., Coppola, G. *et al.* (2004) Testing silicon MEMS structures subjected to thermal loading by digital holography. *Proc. SPIE*, **5343**, 235–243.
30. Ostasevicius, V., Palevicius, A., Daugela, A. *et al.* (2004) Holographic imaging technique for characterization of MEMS switch dynamics. *Proc. SPIE*, **5389**, 73–84.
31. Xu, L., Peng, X.Y., Miao, J.M., and Asundi, A.K. (2001) Studies of digital microscopic holography with applications to microstructure testing. *Appl. Opt.*, **40** (28), 5046–5051.
32. Qin, C., Zhao, J.L., Di, J.L. *et al.* (2009) Visually testing the dynamic character of a blazed-angle adjustable grating by digital holographic microscopy. *Appl. Opt.*, **48** (5), 919–923.
33. Zhao, J.L., Li, E.P., Sun, W.W., and Di, J.L. (2009) Applications of digital holography in visualized measurement of acoustic and flow fields. *Proc. SPIE*, **7522**, 7522221-1-8.
34. Zhang, P., Zhao, J.L., Yang, D.X. *et al.* (2003) Optical masks prepared by using liquid crystal light valve for light-induced photorefractive waveguides. *Appl. Opt.*, **42** (20), 4208–4211.
35. Zhang, P., Zhao, J.L., Yang, D.X. *et al.* (2003) Optically induced photorefractive waveguides in KNSBN:Ce crystal. *Opt. Mat.*, **23** (1–2), 299–303.
36. Yang, D.X., Zhao, J.L., Zhang, P. *et al.* (2003) The index changes of waveguides fabricated by light irradiation in LiNbO3: Fe crystals. *Acta Physica Sinica*, **52** (5), 1179–1183.
37. Zhang, P., Yang, D.X., Zhao, J.L. *et al.* (2004) Light-induced array of three-dimensional waveguides in lithium niobate employing two-beam interference field. *Chin. Phys. Lett.*, **21** (8), 1558–1561.
38. Zhang, P., Zhao, J.L., Yang, D.S. *et al.* (2006) Light-induced waveguide arrays in photorefractive crystals. *Proc. SPIE*, **6032, 60320E-** 1-9.
39. Brandt, E.H. (1989) Levitation in physics. *Science*, **243** (4889), 349–355.
40. Xie, W.J. and Wei, B.B. (2002) Dependence of acoustic levitation capabilities on geometric parameters. *Phys. Rev. E.*, **66** (2), 026605-1-11.
41. Zhang, L., Li, E.P., Feng, W. *et al.* (2005) A study of acoustic levitation process based on laser holographic interferometry. *Acta Phys. Sin.*, **54** (5), 2038–2042. (in Chinese).
42. Zheng, P.C., Li, E.P., Zhao, J.L. *et al.* (2009) Visual measurement of the acoustic levitation field based on digital holography with phase multiplication. *Opt. Commun.*, **282** (22), 4339–4344.

43. Feng, W., Li, E.P., Fan, Q. *et al.* (2005) Diagnosing microwave plasma thruster's plume by digital holographic interferometry. *Acta Photon. Sin.*, **34** (12), 1833–1836. (in Chinese).
44. Tanda, G. and Devia, F. (1998) Application of a schlieren technique to heat transfer measurements in free-convection. *Exp. Fluids*, **24** (4), 285–290.
45. Trainoff, S.P. and Cannell, D.S. (2002) Physical optics treatment of the shadowgraph. *Phys. Fluids*, **14** (4), 1340–1363.
46. Crimaldi, J.P. and Koseff, J.R. (2001) High-resolution measurements of the spatial and temporal scalar structure of a turbulent plume. *Exp. Fluids*, **31** (1), 90–102.
47. Pan, G. and Meng, H. (2003) Digital holography of particle fields: reconstruction by use of complex amplitude. *Appl. Opt.*, **42** (5), 827–833.
48. Herman, C. and Kang, E. (2001) Experimental visualization of temperature fields and study of heat transfer enhancement in oscillatory flow in a grooved channel. *Heat Mass Transfer.*, **37** (1), 87–99.
49. Katti, V. and Prabhu, S.V. (2008) Heat transfer enhancement on a flat surface with axisymmetric detached ribs by normal impingement of circular air jet. *Int. J. Heat Fluid Flow*, **29** (5), 1279–1294.
50. Hossain, M.M. and Shakher, C. (2009) Temperature measurement in laminar free convective flow using digital holography. *Appl. Opt.*, **48** (10), 1869–1877.
51. Colombani, J. and Bert, J. (2007) Holographic interferometry for the study of liquids. *J. Mol. Liq.*, **134** (1–3), 8–14.
52. Gopalan, B. and Katz, J. (2010) Turbulent shearing of crude oil mixed with dispersants generates long microthreads and microdroplets. *Phys. Rev. Lett.*, **104** (5), 054501-1-4.
53. Sun, W.W., Zhao, J.L., Di, J.L. *et al.* (2009) Real-time visualization of Karman vortex street in water flow field by using digital holography. *Opt. Express*, **17** (22), 20342–20348.

Conclusion

Anand Asundi
School of Mechanical and Aerospace Engineering, Nanyang Technological University, Singapore

Digital holography (DH) has brought the field of optical holography a step nearer to practical realization for precision measurement on the micro and nano-scales. Despite the low resolution of the digital (CCD or CMOS) recording media as compared to the high resolution films, the advantage gained by numerical reconstruction of the holograms provides wide-ranging applications in all aspects of mechanics and measurement. Thus, while engineers felt that the holography method was too cumbersome and restricted to lab usage, they now feel that this technique provides a useful advance as devices and structures are shrinking. Digital holography comes in various configurations driven primarily by the application at hand.

To reflect objects such as found in most MEMS and microsystems which use Si-based processes, DH is a tool which can be used throughout the various stages of processing, manufacture and final assembly and testing. During the processing stage, advantage can be taken of the compact designs offered by some lens-less digital holographic systems which allow measurements to be made on-line during the fabrication stage. It is envisaged that these systems might be incorporated into etching as well as thin film coating chambers to monitor the etch depth or coating thickness in real time during the process. This might lead to more reliable and consistent devices. Second, during the device assembly, since different components are packaged together at high temperature and pressures, stress can cause warpage of the device which up

Digital Holography for MEMS and Microsystem Metrology, First Edition. Edited by Anand Asundi.

to now only was manifest when the device did not perform as expected. With DH it is possible to monitor the device during the assembly to see which parts of it are causing issues which can then be corrected. Finally, after the device has been fabricated, putting it through its paces under both static as well as dynamic loading conditions is necessary. DH can help measure and quantify these deformations. Indeed, DH can be used to calibrate devices, including nano-positioners, as well as other sensors and devices. One of the challenges here is the diffraction-limited spatial resolution of holography or for that matter any optical system. Attempts are underway to improve the spatial resolution and one possible approach is discussed in Chapter 5.1.

To transmit objects, the primary applications are the bio-imaging area where single cells can be analyzed and tested. However, for a microlens, both single and array testing are also developing and are a potential area for application as shown in Chapter 3. Challenges to lens testing are similar to what are observed using conventional interferometers but the ability to measure small lenses with a very compact and simple geometry offers the greatest advantage of DH. Other applications involve measuring parameters such as temperature, or fluid flow where there are changes in the refractive index caused due to these parameters. Using double exposure holographic interferometry, small variations in the refractive index can be readily monitored. However, one of the challenges for transparent objects is that the phase change is caused due to the change in thickness as well as a change in refractive index. However, from a single hologram, it would not be possible to separate the two effects – hence alternatives are needed. Approaches such as tomography might offer some solution to this problem.

Digital holography has enabled the realization of the first holograms that Denis Gabor had suggested were possible. Digital reconstruction allows us to overcome the zero-order and twin image problems which had plagued in-line holograms. Thus, particle imaging and measurement become a reality. Unlike image-based systems, digital holography allows the recording of the volume and then isolating planes within it. Hence one could think of DH as a system recording first and focusing afterwards, as opposed to conventional imaging which requires one to focus first and record later. Particle imaging applications are shown for the crystallization process. However, there are wider applications of this simple set-up in pollution monitoring, bubble generator characterization and underwater imaging.

In conclusion, DH has now become quite common and the technology is poised to move out of the labs and into industry. This should enable 3D imaging and inspection to be taken to the next level. Having said that, challenges in various sectors still need to be overcome but this can be done with innovative optical designs and enhanced reconstruction algorithms.

Index

Digital Holography for MEMS and Microsystem Metrology, First Edition. Edited by Anand Asundi.

CPSIA information can be obtained at www.ICGtesting.com
Printed in the USA
267299BV00001B/4/P

9 780470 978696